Type Specimens of Fossil Fishes

Catalogue of the University of Alberta Laboratory for Vertebrate Paleontology

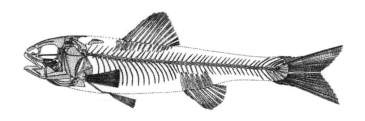

†Massamorichthys wilsoni Murray, 1996

Type Specimens of Fossil Fishes

Catalogue of the University of Alberta Laboratory for Vertebrate Paleontology

John Clay Bruner
Department of Biological Sciences
University of Alberta
Edmonton, Alberta
T6G 2E9 Canada

CRC Press
Taylor & Francis Group
Boca Raton London New York

CRC Press is an imprint of the
Taylor & Francis Group, an **informa** business

Figure on cover is Figure 2. Reconstruction of †*Massamorichthys wilsoni*, gen. et sp. nov., page 644, from the following article: Alison M. Murray. 1996. A new Paleocene genus and species of percopsid, †*Massamorichthys wilsoni* (Paracanthopterygii) from Joffre Bridge, Alberta, Canada. *Journal of Vertebrate Paleontology* Vol. 16(4):642–652.

DOI: http://dx.doi.org/10.1080/02724634.1996.10011354

Reprinted by permission of the Society of Vertebrate Paleontology

CRC Press
Taylor & Francis Group
6000 Broken Sound Parkway NW, Suite 300
Boca Raton, FL 33487-2742

© 2019 by Taylor & Francis Group, LLC
CRC Press is an imprint of Taylor & Francis Group, an Informa business

No claim to original U.S. Government works

Printed on acid-free paper

International Standard Book Number-13: 978-0-367-02642-4 (Hardback)

Visit the Taylor & Francis Web site at
http://www.taylorandfrancis.com

and the CRC Press Web site at
http://www.crcpress.com

Contents

Introduction

The institutional responsibilities regarding type specimens, under Recommendation 72D of the International Code of Zoological Nomenclature, were discussed by Berry (1985, p. 25) as follows: "Every institution in which types are deposited should (1) ensure that all are clearly marked so that they will be unmistakably recognized; (2) take all necessary steps for their safe preservation; (3) make them accessible for study; (4) publish lists of type material in its possession or custody; and (5) so far as is possible, communicate information concerning types when requested by zoologists." According to Langston *et al.* (1977, p. 7), "...the holotype is the namebearer for a species and in practical terms constitutes a standard of comparison and recognition for that species. Under the Code there is a quasi-legal obligation on the part of any institution possessing such a specimen to take all necessary steps for its safe and perpetual preservation."

While moving the University of Alberta Laboratory for Vertebrate Paleontology (UALVP) reference collection in the cabinets on the compactors from the old CW-007 room to the new paleontology collection room in CCIS L2-245, the UALVP curators requested that the types be separated from the reference collections and stored in lockable type cabinets. However, the UALVP curators had no idea as to how many types were in their collections. A search made of the computer catalogue in January 2017 for the fish collection listed a total of 159 holotypes and 781 paratypes of fishes. The actual number of fish types found in the

UALVP collection as of October 1, 2018, was 89 holotypes and 978 paratypes and 56 casts of holotypes and 23 casts of paratypes from other museums.

This type catalogue represents the first time the UALVP has had a complete listing of the type fossil fishes deposited in its collections. The UALVP had no accurate idea as to how many fish types were in their collections. In some cases, specimen lots were never marked as type specimens, and in other cases, specimens marked as types were never published. This list corrects mistakes made by the original authors in citing the type specimens. Some authors cited the wrong catalogue numbers or were given already used catalogue numbers. In some cases, the same catalogue number for type specimens had been used as many as three times for other specimens from other geologic ages and classes. When this discovery was made, whoever published first was given the original UALVP catalogue number, and the other specimens sharing that number were given a new catalogue number but linked to the old number. Also, the same specimens have been used for as many as three different type series as more information became available and more correct determinations could be made. It is hoped that this type catalogue will lead to the discovery of more fossil type specimens deposited in the UALVP which have been left unmarked and forgotten. The specimens in this type list also include specimens identified as new in MSc and PhD theses although they are not yet published. These unpublished types are labeled as types and paratypes in the computer catalogue.

I have also included a listing of casts of type specimens from other museums in the fish collections. Often a cast can be as useful as the original specimen. Normally workers would not think of checking for copies of types at a museum other than the one at which the original is deposited. And, prior to this catalogue, no list of casts of type specimens from other museums in the UALVP had ever been prepared. Some museums will not lend their type specimens. This list may be useful to those workers who need to see a specimen but may not be able to travel to the museum where the original is deposited. The fish collection contains 56 casts of holotypes and 23 casts of paratypes.

This catalogue provides the latest classification if the original identification of the type is not considered valid. The classification scheme followed is a combination of Bruner (1992), Cappetta (1987), Denison (1978), Grande and Bemis (1998), Märss et al. (2007), Nelson et al. (2016), and Zangerl (1981). The species' names are spelled exactly as they appear in the original description. This listing gives the catalogue number (in some cases, multiple numbers have been assigned to the same lot). It gives a brief description of the specimen. The geologic age and the type locality are given for the primary types. The names of the collectors are given of the holotype if known. Also, all paratype specimens in the UALVP's fish collections are given. When citing fossil fishes in the UALVP collections, the abbreviation UALVP should be given followed directly by the catalogue number without a space in between. Also, the current resting place in the reference collections is given if known. For example, "Silurian 11, 02" means the specimen is in the Silurian cabinet 11, in shelf 2.

The following authors have described 74 holotypes (H) and 821 paratypes (P) in the UALVP fish collection: Adrain and Wilson (1994) 2 H, 22 P; Blais, Hermus, and Wilson (2015) 4H, 25 P; Case, Cook, Sadorf, and Shannon (2017) 1 H, 1 P; Case, Cook, and Wilson (2011) 1 H, 4 P; Case, Cook, Wilson, and Borodin (2012) 1 H, 3 P; Cook, Wilson, Murray, Plint, Newbrey, and Everhart (2013) 1 H, 3 P; Forey (1975) 1 H, 71 P; Gagnier, Hanke, and Wilson (1999) 1 H, 5 P; Gagnier and Wilson (1996) 1 H; Hanke (2008) 1 H, 6 P; Hanke, Davis, and Wilson (2001) 1 H, 6 P; Hanke and Wilson (2004) 2 H, 20 P; Hanke, Wilson, and Lindoe (2001) 2 H, 5 P; Li and Wilson (1994) 1 H, 1 P; Li and Wilson (1996) 1 H, 17 P; Liu, Wilson, and Murray (2016) 1 H, 245 P; Märss (1999) 1 H, 1 P; Märss and Gagnier (2001) 1 H, 22 P; Märss, Wilson, and Thorsteinsson (2002) 16 H, 4 P; Murray (1996) 1 H, 29 P; Murray (2016) 1 H; Murray and Cumbaa (2013) 1 H; Murray and Wilson (1996) 1 H, 3 P; Murray and Wilson (2009) 1

H; Murray and Wilson (2011) 1 H, 2 P; Murray and Wilson (2013) 2 H, 13 P; Murray and Wilson (2014) 3 H, 5 P; Mutter, De Blanger, and Neuman (2007) 1 H, 4 P; Mutter and Neuman (2006) 1 H, 70 P; Mutter and Neuman (2008) 1 H; Neuman and Mutter (2005) 1 H, 62 P; Russell (1928) 1 H, 2 P; Scott and Wilson (2014) 1 H, 18 P; Soehn and Wilson (1990) 1 H, 12 P; Soehn, Märss, Caldwell, and Wilson (2001) 2 H, 31 P; Thomson (1967) 1 H; Vernygora, Murray, and Wilson (2016) 1 H; Wendruff and Wilson (2013) 1 H; Wilson (1978) 1 H, 6 P; Wilson (1979) 1 H, 2 P; Wilson (1980) 1 H, 5 P; Wilson (1982) 1 H, 5 P; Wilson and Murray (1996) 1 H, 13 P; Wilson and Caldwell (1998) 5 H, 34 P; and Wilson and Williams (1991) 1 H, 44 P.

Casts of 56 holotypes (CH) and casts of 23 paratypes (CP) in the UALVP fish collection have been described by Applegate (1988) 1 CH; Applegate (1992) 1 CH; Cavender (1969) 1 CH, 2 CP; Claypole (1885) 1 CH; Cope (1872) 1 CH; Daeschler, Shubin, and Jenkins, Jr. (2006) 1 CH; Denison (1953) 1 CH; Denison (1963) 7 CH, 1 CP; Denison (1964) 6 CH, 6 CP; Dineley (1976) 1 CH, 2 CP; Dineley and Loeffler (1976) 16 CH, 12 CP; Flower and Wayland-Smith (1952) 3 CH; Fox, Campbell, Barwick, and Long (1995) 1 CH; Geinitz (1884) 1 CH; Grande (1979) 1 CH; Hanke, Stewart, and Lammers (1996) 1 CH; Hanke, Stewart, and Lammers (1996) 1 CH; Hussakof (1916) 1 CH; Kiaer (1932) 4 CH; Matthew (1888) 1 CH; Murray, Simons, and Attia (2005) 1 CH; Schaeffer (1949) 1 CH; Stensiö (1958) 1 CH; Stetson (1931) 1 CH; and Traquair (1898) 1 CH.

Blais and Wilson have described two holotypes (SH) in the UALVP fish collection in a submitted publication.

The following authors described 13 holotypes (H MSc, H PhD) and 156 paratypes (P MSc, P PhD) in MSc and PhD theses in the UALVP fish collection: Greeniaus (2004) 4 H MSc, 15 P MSc; Hanke (2001) 4 H PhD, 132 P PhD; Hawthorn (2009) 4 H MSc, 2 P MSc; and Wendruff (2011) 1 H MSc, 8 P MSc.

The earliest collected fish holotype catalogued into the UALVP collections is UALVP131 *Cyclurus lacus* (Russell, 1928), Collector: Loris S. Russell Sept. 1924. Canada: Alberta, Alberta, Red Deer River, Red Deer River #13. Tertiary-Paleocene, Paskapoo Formation. Partial Left splenial with grinding teeth.

Type Specimens in the Collections of the University of Alberta Laboratory for Vertebrate Paleontology

Agnatha Haeckel, 1895
   Heterostraci Lankester, 1868
      Incertae Familiae
UALVP34240 Cast of HOLOTYPE, NMC 21460 at National Museum of Canada. *Aserotaspis canadensis* **Dineley and Loeffler, 1976, p. 137**
Canada: Northwest Territories. Silurian-Silurian Late-Ludlovian? Delorme Group? Variation in ornamentation of dorsal surface of cephalothorax. Dineley and Loeffler, 1976, Descr. pp. 136–140, & fig. Plate 18, p. 139, Text-Fig. 57, p. 137. CCIS L2-245, Silurian 10, 08

Dineley, D. L., and Loeffler, E. J. 1976. Ostracoderm faunas of the Delorme and associated Siluro-Devonian formations North West Territories, Canada. Special Papers in Palaeontology. No. 18:1–214.

Agnatha Haeckel, 1895
  Heterostraci Lankester, 1868
    Cyathaspidiformes Kiaer and Heintz, 1935
      Cyathaspidida Tarlo, 1962
        Cyathaspididae Kiaer, 1932
UALVP40742 Cast of HOLOTYPE Cast of FMNH PF 737 at the Field Museum of Natural History *Allocryptaspis utahensis* Denison, 1953, p. 294
Collectors: Denison, Robert H., Turnbull, William Davey, and Turnbull, Priscilla F., July 1950.
The United States: Utah: Cache County: Cottonwood Canyon, NW1/4, Sec. 18, T13N, R3E, Site 6. Early Devonian: Water Canyon Fm. Dorsal and ventral shields, associated body scales. Denison, 1953, Descr. pp. 294–304, & fig. Fig. 61, p. 295, Fig. 62, p. 297. CCIS L2-245, Devonian 8, 12
  Denison, Robert H. 1953. Early Devonian fishes from Utah. Part II. Heterostraci. Fieldiana: Geology Vol. 11(7):291–355.

Agnatha Haeckel, 1895
  Heterostraci Lankester, 1868
    Cyathaspidiformes Kiaer and Heintz, 1935
      Cyathaspidida Tarlo, 1962
        Cyathaspididae Kiaer, 1932
UALVP40712 Cast of HOLOTYPE, FMNH PF 866. *Americaspis claypolei* Denison, 1964, p. 416
Collectors: Denison, Robert H., and Denison, L. B., August 1951.
The United States: New York, Orange County; Greenville Twp., Shin Hollow along Erie Railroad tracks, about 1.75 miles SSE of Graham Station, Locality La. Late Silurian, High Falls Fm.,

Longwood Shale. Dorsal shield. Denison, 1964, Descr. pp. 416–419, Fig. 141, A, B, p. 417 CCIS L2-245, Silurian 11, 02

    Denison, Robert H. 1964. The Cyathaspididae. A family of Silurian and Devonian jawless vertebrates. Fieldiana: Geology Vol. 13(5):311–473.

Agnatha Haeckel, 1895
    Heterostraci Lankester, 1868
        Cyathaspidiformes Kiaer and Heintz, 1935
            Cyathaspidida Tarlo, 1962
            Cyathaspididae Kiaer, 1932

UALVP40745 Cast of HOLOTYPE Cast of FMNH PF 3672, original at National Museum of Canada, NMC 10038. *Anglaspis expatriata* **Denison, 1964, p. 431**

Canada: Northwest Territories: beside Keele River, Top of Mount Sekwi, 63°28'N. Lat., 128°40'W. Long., California Standard Locality Z 29–61, 4474'. Early Devonian. Cast of Dorsal shield. Denison, 1964, Descr. pp. 431–433, & fig. Fig. 151, A, p. 432. CCIS L2-245, Silurian 8, 12

    Denison, Robert H. 1964. The Cyathaspididae. A family of Silurian and Devonian jawless vertebrates. Fieldiana: Geology Vol. 13(5):311–473.

Agnatha Haeckel, 1895
    Heterostraci Lankester, 1868
        Cyathaspidiformes Kiaer and Heintz, 1935
            Cyathaspidida Tarlo, 1962
            Cyathaspididae Kiaer, 1932

UALVP40711 Cast of HOLOTYPE Cast of FMNH PF 1132 which is a cast of PMOD 186 *Anglaspis insignis* **Kiaer, 1932**

Norway: Spitsbergen: Red Bay, Fraenkelryggen. Early Devonian: *Anglaspis* Horizon, Red Bay Series, Fraenkelryggen Group. Five shields on slab. Kiaer, 1932, Descr. p. 20, & Plate VII, Figs. 1, 2. CCIS L2-245, Silurian 8, 12

Kiaer, Johan. 1932. The Downtonian and Devonian vertebrates of Spitsbergen. IV. Suborder Cyathaspida. Skrifter om Svalbard og Ishavet. 52:1–26.

Agnatha Haeckel, 1895
  Heterostraci Lankester, 1868
    Cyathaspidiformes Kiaer and Heintz, 1935
      Cyathaspidida Tarlo, 1962
        Cyathaspididae Kiaer, 1932
UALVP40746 Cast of HOLOTYPE Cast of FMNH PF 3266, original at Museum of Comparative Zoology *Archegonaspis drummondi* Flower and Wayland-Smith, 1952, p. 380
The United States: New York: Oneida County: about 2 mi. SE of Kenwood. Late Silurian: Salina Group, Vernon Shale. Cast of inner surface of dorsal shield. Flower and Wayland-Smith, 1952, Descr. pp. 380–381, & fig. Plate 2, Figs. 1, 2. CCIS L2-245, Silurian 11, 02
  Flower, R. H., and Wayland-Smith, R. 1952. Cyathaspid fishes from the Vernon Shale of New York. Bulletin of the Museum of Comparative Zoology Vol. 107(6):355–387.

Agnatha Haeckel, 1895
  Heterostraci Lankester, 1868
    Cyathaspidiformes Kiaer and Heintz, 1935
      Cyathaspidida Tarlo, 1962
        Cyathaspididae Kiaer, 1932
UALVP40728 Cast of HOLOTYPE Originally described as *Cyathaspis schmidti* Geinitiz, 1884. Cast of FMNH PF 1128, original at Oslo Paleontology Museum. *Archegonaspis integer schmidti* (Geinitiz, 1884)
Germany: Bahnhof: Rostock. Late Silurian: Graptolithengestein. Ventral shield. Geinitz, 1884, Descr. pp. 854–857. (=*Archegonaspis integer schmidti* (Geinitz, 1884)) CCIS L2-245, Silurian 11, 02
  Geinitz, F. E. 1884. Ueber ein Graptolithenführendes Geschibe mit *Cyathaspis* von Rostock. Zeitschrift der Deutschen geologischen Gesellschaft Vol. 36(4):854–857.

Agnatha Haeckel, 1895
   Heterostraci Lankester, 1868
      Cyathaspidiformes Kiaer and Heintz, 1935
         Cyathaspidida Tarlo, 1962
            Cyathaspididae Kiaer, 1932; or, Ariaspididae
UALVP34188 and UALVP40748 2 Casts of HOLOTYPE PU 17103 at Princeton University Geological Museum *Ariaspis ornata* **Denison, 1963, pp. 120–123**
Collectors: Bamber, E. W., and MacDonald, W. D., 1959.
Canada: SE Yukon Territory, Beaver River, Silurian-Silurian Late-Ludlovian? Delorme Group? Dorsal shield, and posterior part of ventral shield. Denison, 1963, Descr. pp. 120–123, & fig. Fig. 68, p. 121, Fig. 69, p. 122. UALVP34188 in CCIS L2-245, Silurian 11, 06; UALVP40748 in CCIS L2-245, Silurian 11, 02
   Denison, Robert H. 1963. New Silurian Heterostraci from southeastern Yukon. Fieldiana: Geology, 14(7):105–141.

Agnatha Haeckel, 1895
   Heterostraci Lankester, 1868
      Cyathaspidiformes Kiaer and Heintz, 1935
         Cyathaspidida Tarlo, 1962
            Cyathaspididae Kiaer, 1932
UALVP15682 HOLOTYPE. *Athenaegis chattertoni* **Soehn and Wilson, 1990, p. 487**
Collector: Chatterton, Brian D. E., 1978
Canada: Northwest Territories, Avalanche Lake, AV1, Silurian-Silurian Early-Wenlock, Delorme Fm. Dorsal view complete. Soehn and Wilson, 1990, Descr. p. 407, & fig. Fig. 4, p. 411, Fig. 5, C, E, p. 412; Table 1, p. 407. Earth Sciences Museum, ESB B-01; casts in teaching collection L1/L2
   Soehn, Kenneth L. and Wilson, Mark V. H. 1990. A complete, articulated Heterostracan from Wenlockian (Silurian) beds of the Delorme Group, Mackenzie Mountains, Northwest Territories, Canada. Journal of Vertebrate Paleontology Vol. 10(4):405–419

PARATYPES:

UALVP15679 head shield with part of trunk. Soehn and Wilson, 1990, Descr. p. 407, & fig. Fig. 2, C, p. 409; Table 1, p. 407. CCIS L2-245, Silurian 05, 10

UALVP15680 head shield, articulated anterior half of an individual with ventral shield and crushed oral area exposed. Soehn and Wilson, 1990, Descr. p. 407. CCIS L2-245, Silurian 05, 10

UALVP15681 head shield with part of trunk. Soehn and Wilson, 1990, Descr. p. 407, & fig. Fig. 2, A, p. 409; Fig. 5, F, p. 412; Table 1, p. 407. CCIS L2-245, Silurian 05, 10

UALVP15683 partially covering part of head shield and trunk. Soehn and Wilson, 1990, Descr. p. 407, & fig. Fig. 2, B, G, H, I, p. 409; Fig. 3, p. 410; Table 1, p. 407. Earth Sciences Museum, ESB B-01; 2 casts in teaching collection L1/L2 Z-425

UALVP28216 fish skeleton. Soehn and Wilson, 1990, Descr. p. 407. CCIS L2-245, Silurian 05, 10

UALVP29663 head shield with part of trunk. Soehn and Wilson, 1990, Descr. p. 407, & fig. Fig. 3, p. 410.

UALVP29664 head shield with part of trunk. Soehn and Wilson, 1990, Descr. p. 407, & fig. Fig. 2, D, E, p. 409; Fig. 6, B, C, p. 413. CCIS L2-245, Silurian 05, 10

UALVP29665 single partial specimen. Soehn and Wilson, 1990, Descr. p. 407. CCIS L2-245, Silurian 05, 10

UALVP29666 single partial specimen. Soehn and Wilson, 1990, Descr. p. 407. CCIS L2-245, Silurian 05, 10

UALVP29667 single partial specimen. Soehn and Wilson, 1990, Descr. p. 407, & fig. Fig. 2, F, p. 409. CCIS L2-245, Silurian 05, 10

UALVP29668 single partial specimen. Soehn and Wilson, 1990, Descr. p. 407, & fig. Fig. 2, A, p. 409, Fig. 5B, p. 412. CCIS L2-245, Silurian 05, 10

UALVP29926 ventral shield and articulated mouth parts, some body scales. Soehn and Wilson, 1990, Descr. p. 407, & fig. Fig. 5G, p. 412

Agnatha Haeckel, 1895
 Heterostraci Lankester, 1868

Cyathaspidiformes Kiaer and Heintz, 1935
Cyathaspidida Tarlo, 1962
Cyathaspididae Kiaer, 1932
UALVP40732 Cast of HOLOTYPE Cast of FMNH PF 3777 which is a cast of the Royal Ontario Museum ROM 1117. UALVP34269 cast of New Brunswick Museum of Geology NBMG3072. Cast of HOLOTYPE of *Diplaspis acadica* Matthew, 1888. *Cyathaspis acadica* (Matthew, 1888)
Canada: New Brunswick: Kings County: Nerepis River. Silurian. Plates. Matthew, 1888, Descr. pp. 49–62, & fig. Plate 4. UALVP40732 CCIS L2-245, Silurian 11, 02 UALVP34269 CCIS L2-245, Silurian 10, 09

Matthew, G. F. 1888. On some remarkable organisms of the Silurian and Devonian rocks in southern New Brunswick. Royal Society of Canada, Proceedings and Transcripts Vol. 6(1) Sec. 4:49–62.

Agnatha Haeckel, 1895
  Heterostraci Lankester, 1868
    Cyathaspidiformes Kiaer and Heintz, 1935
      Cyathaspidida Tarlo, 1962
        Cyathaspididae Kiaer, 1932
UALVP40716 Cast of HOLOTYPE, Oslo Paleontological Museum, PMOD 454, *Dinaspidella robusta* (Kiaer, 1932, p. 18), Originally described as *Dinaspis robusta* Kiaer, 1932
Spitsbergen: Fraenkelryggen: Red Bay. Early Devonian: Red Bay Series: Fraenkelryggen Group, Primaeva Horizon. Kiaer, 1932, Descr. p. 18, & fig. Figs. 7, 8, p. 16, and Plate IV, Figs. 2, 3. CCIS L2-245, Devonian 8, 12

Kiaer, Johan. 1932. The Downtonian and Devonian vertebrates of Spitsbergen. IV. Suborder Cyathaspida. Skrifter om Svalbard og Ishavet. 52:1–26.

Agnatha Haeckel, 1895
  Heterostraci Lankester, 1868

Cyathaspidiformes Kiaer and Heintz, 1935
Cyathaspidida Tarlo, 1962
Cyathaspididae Kiaer, 1932
UALVP34189 Cast of HOLOTYPE Cast of HOLOTYPE
PU 17101 at Princeton University Geological Museum.
Homalaspidella borealis Denison, 1963, pp. 123–127
Collectors: Bamber, E. W., and MacDonald, W. D., 1959.
Canada: SE Yukon Territory, Beaver River, Silurian-Silurian
Late-Ludlovian? Delorme Group? Dorsal shield. Denison, 1963,
Descr. pp. 123–127, & fig. Fig. 70, p. 124, Fig. 71, p. 125, Fig. 72,
p. 126. CCIS L2-245, Silurian 11, 06
   Denison, Robert H. 1963. New Silurian Heterostraci from
southeastern Yukon. Fieldiana: Geology Vol. 14(7):105–141.
Cast of PARATYPE:
UALVP34190 cast of paratype at Princeton University Geological
Museum

Agnatha Haeckel, 1895
   Heterostraci Lankester, 1868
      Cyathaspidiformes Kiaer and Heintz, 1935
         Cyathaspidida Tarlo, 1962
            Cyathaspididae Kiaer, 1932
UALVP40725 Cast of HOLOTYPE, Oslo Paleontological
Museum, PMOD 156, *Homalaspidella nitida* (Kiaer, 1932),
Originally described as *Homalaspis nitida* Kiaer, 1932
Spitsbergen: Red Bay. Early Devonian: Red Bay Series, Horizon A.
Shield Kiaer, 1932, Descr. pp. 14–25, & fig. Fig. 6, p. 15, Plate IV,
Fig. 1. CCIS L2-245, Devonian 8, 12
   Kiaer, Johan. 1932. The Downtonian and Devonian ver-
tebrates of Spitsbergen. IV. Suborder Cyathaspida. Skrifter om
Svalbard og Ishavet. 52:1–26.

Agnatha Haeckel, 1895
   Heterostraci Lankester, 1868

Cyathaspidiformes Kiaer and Heintz, 1935
Cyathaspididae Kiaer, 1932
UALVP40743 Cast of HOLOTYPE Cast of FMNH PF 3694,
original at National Museum of Canada, NMC 10030. *Listraspis*
canadensis **Denison, 1964, pp. 386–388**
Canada: British Columbia: about 30 miles NW of Muncho Lake,
on Alaska Highway, mile 450, 59°7'30"N. Lat., 126°22'W. Long.,
about 30 miles NW of Muncho Lake, Locality R5–61, 2100' of
California Standard company. Early Devonian. Cast of dorsal
shield. Denison, 1964, Descr. pp. 391–396, & fig. Fig. 131, p. 391.
CCIS L2-245, Devonian 8, 12

Denison, Robert H. 1964. The Cyathaspididae. A family of
Silurian and Devonian jawless vertebrates. Fieldiana: Geology
Vol. 13(5):309–473.

<u>PARATYPE:</u>
UALVP40755 cast of FMNH PF 3682 at Field Museum of Natural
History. Cast of incomplete dorsal shield. Denison, 1964, Descr.
pp. 391, 395, & fig. Fig. 132, p. 393. CCIS L2-245, Devonian 8, 12

Agnatha Haeckel, 1895
　　Heterostraci Lankester, 1868
　　　　Cyathaspidiformes Kiaer and Heintz, 1935
　　　　　　Cyathaspidida Tarlo, 1962
　　　　　　　　Cyathaspididae Kiaer, 1932
UALVP34250 Cast of HOLOTYPE Cast of NMC 19819,
Original at Museum of Canada. *Natlaspis planicosta* **Dineley**
and Loeffler, 1976, p. 140
Canada: Northwest Territories. Silurian-Silurian Late-Ludlovian,
Delorme Group. Cast of incomplete dorsal shield. Dineley and
Loeffler, 1976, Descr. pp. 140–144, & fig. Plate 19, Fig. 1, p. 143,
Text-Fig. 58, p. 141. CCIS L2-245, Silurian 10, 9

Dineley, D. L., and Loeffler, E. J. 1976 Ostracoderm faunas of
the Delorme and associated Siluro-Devonian formations North West
Territories, Canada. Special Papers in Palaeontology, No. 18:1–214.

Agnatha Haeckel, 1895
 Heterostraci Lankester, 1868
  Cyathaspidiformes Kiaer and Heintz, 1935
   Cyathaspidida Tarlo, 1962
    Cyathaspididae Kiaer, 1932
UALVP40730 Cast of HOLOTYPE. Cast of FMNH PF 3703 which is a cast of NMC 10034, original at National Museum of Canada. *Pionaspis acuticosta* Denison, 1964.
Canada: British Columbia: about 30 miles NW of Muncho Lake, on Alaska Highway, Mile 450, 59°7′30″N. Lat., 126°22′W. Long., Locality R5–61, 2100′ of California Standard Company. Early Devonian. Incomplete dorsal shield. Denison, 1964, Descr. pp. 388–390, & fig. Fig. 129, p. 388, Fig. 130, A, B, p. 389. CCIS L2-245, Devonian 8, 12

Denison, Robert H. 1964. The Cyathaspididae. A family of Silurian and Devonian jawless vertebrates. Fieldiana: Geology Vol. 13(5):309–473.

Agnatha Haeckel, 1895
 Heterostraci Lankester, 1868
  Cyathaspidiformes Kiaer and Heintz, 1935
   Cyathaspidida Tarlo, 1962
    Cyathaspididae Kiaer, 1932
UALVP34239 Cast of HOLOTYPE. Cast of NMC 19640. *Pionaspis amplissima* Dineley and Loeffler, 1976, p. 64
Canada: Northwest Territories. Silurian-Silurian Late-Ludlovian? Delorme Group?, Dorsal shield. Dineley and Loeffler, 1976, Descr. pp. 64–70, & fig. Plate 8, Fig. 1, Fig. 2, p. 67, Plate 9, p., Text-Fig. 24, p. 68, Text-Fig. 37, p. 105. CCIS L2-245, Silurian 10, 08

Dineley, D. L., and Loeffler, E. J. 1976. Ostracoderm faunas of the Delorme and associated Siluro-Devonian formations North West Territories, Canada Special Papers in Palaeontology, No. 18:1–214.

PARATYPES:
UALVP34243 cast of NMC 12888, lateral scale. Dineley and Loeffler, 1976, Descr. pp. 64–70, & fig. Plate 9, Fig. 5, p. 69. CCIS L2-245, Silurian 10, 08
UALVP34244 cast of NMC 12818, fulceral scale. Dineley and Loeffler, 1976, Descr. pp. 64–70, & fig. Plate 9, Fig. 4, p. 69. CCIS L2-245, Silurian 10, 09
UALVP34245 cast of NMC 19611, dorsal ridge scale. Dineley and Loeffler, 1976, No. 18:1–214, Descr. pp. 64–70, & fig. Plate 9, Fig. 3, p. 69. CCIS L2-245, Silurian 10, 09
UALVP34246 cast of NMC 19682, incomplete ventral shield. Dineley and Loeffler, 1976, Descr. pp. 64–70, & fig. Plate 9, Fig. 1, p. 69. CCIS L2-245, Silurian 10, 09

Agnatha Haeckel, 1895
   Heterostraci Lankester, 1868
      Cyathaspidiformes Kiaer and Heintz, 1935
         Cyathaspidida Tarlo, 1962
            Cyathaspididae Kiaer, 1932
UALVP40734 Cast of HOLOTYPE Cast of FMNH PF 3702 which is a cast of NMC 10035, at the National Museum of Canada. *Pionaspis planicosta* **Denison, 1964, pp. 386–388**
Canada: British Columbia: about 30 miles NW of Muncho Lake, on Alaska Highway, mile 450, 59°7'30"N. Lat., 126°22'W. Long., Locality R5-61, 2100' of California Standard company. Early Devonian. Dorsal shield. Denison, 1964, Descr. pp. 386–388, & fig. Fig. 126, A, B, p. 386, Fig. 127, p. 387. CCIS L2-245, Devonian 8, 12
   Denison, Robert H. 1964. The Cyathaspididae. A family of Silurian and Devonian jawless vertebrates. Fieldiana: Geology Vol. 13(5):309–473.
PARATYPE:
UALVP40735 cast of ventral shield. Cast of FMNH PF 3681 at Field Museum of Natural History, which is a cast of the original at

the National Museum of Canada. Denison, 1964, Descr. p. 387, & fig. Fig. 128, p. 387. CCIS L2-245, Devonian 8, 12

Agnatha Haeckel, 1895
  Heterostraci Lankester, 1868
    Cyathaspidiformes Kiaer and Heintz, 1935
      Cyathaspidida Tarlo, 1962
        Cyathaspididae Kiaer, 1932
UALVP34186 and UALVP40718 Two casts of HOLOTYPE Two casts of HOLOTYPE PU 17090 at Princeton University Geological Museum *Ptomaspis canadensis* Denison, 1963, pp. 113–116
Collectors: Bamber, E. W., and MacDonald, W. D., 1959. Canada: Southeastern Yukon Territory, Beaver River, 60°27′N. Lat., 125°46′W. Long. Silurian-Silurian Late-Ludlovian? Delorme Group? Nearly complete Dorsal shield. Denison, 1963, Descr. pp. 113–116, & fig. Fig. 64, p. 114, Fig. 65, p. 115. UALVP40718 in CCIS L2-245, Silurian 11, 02; UALVP34186 in CCIS L2-245, Silurian 11, 06
  Denison, Robert H. 1963. New Silurian Heterostraci from southeastern Yukon. Fieldiana: Geology, 14(7):105–141.

Agnatha Haeckel, 1895
  Heterostraci Lankester, 1868
    Cyathaspidiformes Kiaer and Heintz, 1935
      Cyathaspidida Tarlo, 1962
        Cyathaspididae Kiaer, 1932
UALVP40736, UALVP40737 Cast of HOLOTYPE. Cast of FMNH PF 3659 which is a cast of C1618 a, b, at the Naturistoriska Riksmusset. *Seretaspis zychi* Stensiö, 1958, p. 371
Czech Republic: Podolia. Devonian: Downtonian: presumably Czortków Series. Two sets of casts of this specimen, part & counterpart, UALVP 40736 and UALVP 40737. Stensiö, 1958, Descr. pp. 371, 384, 386, 391, 393, & fig. Fig. 204, p. 381. CCIS L2-245, Devonian 8, 12

Stensiö, E. 1958. Les Cyclostomes Fossiles ou Ostracodermes. *IN*: Grassé, P.-P. (editor) Traité de Zoologie Anatomie, Systématique, Biologie, 13 (fasc. 1):173–425.

Agnatha Haeckel, 1895
Heterostraci Lankester, 1868
Cyathaspidiformes Kiaer and Heintz, 1935
Cyathaspidida Tarlo, 1962
Cyathaspididac Kiaer, 1932

UALVP40739 and UALVP40749 Casts of HOLOTYPE Cast of FMNH PF 3262, original at Museum of Comparative Zoology, Harvard University. *Vernonaspis allenae* **Flower and Wayland-Smith, 1952, p. 375**

The United States: New York: Oneida County: about 2 miles SE of Kenwood. Late Silurian: Salina Group, Vernon Shale. Dorsal shield in counterpart. Flower and Wayland-Smith, 1952, Descr. pp. 375–376, & fig. Plate 1, Plate 2, Figs. 3, 8. CCIS L2-245, Silurian 11, 02

Flower, R. H., and Wayland-Smith, R. 1952. Cyathaspid fishes from the Vernon Shale of New York. Bulletin of the Museum of Comparative Zoology Vol. 107(6):355–387.

Agnatha Haeckel, 1895
Heterostraci Lankester, 1868
Cyathaspidiformes Kiaer and Heintz, 1935
Cyathaspidida Tarlo, 1962
Cyathaspididae Kiaer, 1932

UALVP34184 AND UALVP40719 Two casts of HOLOTYPE. 2 Casts of HOLOTYPE PU 17801 at Princeton University Geological Museum. *Vernonaspis bamberi* **Denison, 1963, pp. 108–110**

Collectors: Bamber, E. W., and MacDonald, W. D., 1959.

Canada: Yukon Territory, Beaver River. Silurian-Silurian Late-Ludlovian?, Delorme Group? Dorsal shield, nearly complete. Denison, 1963, Descr. pp. 108–110, & fig. Fig. 60, p. 109, Fig. 61,

p. 110. UALVP34184 in CCIS L2-245, Silurian 11, 06; UALVP40719 in CCIS L2-245, Silurian 11, 02

Denison, Robert H. 1963. New Silurian Heterostraci from southeastern Yukon. Fieldiana: Geology Vol. 14(7):105–141.

Agnatha Haeckel, 1895
  Heterostraci Lankester, 1868
    Cyathaspidiformes Kiaer and Heintz, 1935
      Cyathaspidida Tarlo, 1962
      Cyathaspididae Kiaer, 1932
UALVP40752 Cast of PARATYPE, Cast of PU 12922, original at Princeton University. *Vernonaspis bryanti* **Denison, 1964, p. 377**
Collectors: Robert H. and L. B. Denison
The United States: New Jersey: Sussex County: Montague Twp., grove along New Jersey Highway 23, near Mill Brook, road cut opposite Vie's Lunch Stand, Locality 1, east side of New Jersey Highway 23, 0.9 mi (1.4 mi by road) SSE of Duttonville. Late Silurian: High Falls Fm. Cast of dorsal shield. Denison, 1964, Descr. p. 377. CCIS L2-245, Silurian 11, 02

Denison, Robert H. 1964. The Cyathaspididae. A family of Silurian and Devonian jawless vertebrates. Fieldiana: Geology Vol. 13(5):309–473.

Agnatha Haeckel, 1895
  Heterostraci Lankester, 1868
    Cyathaspidiformes Kiaer and Heintz, 1935
      Cyathaspidida Tarlo, 1962
      Cyathaspididae Kiaer, 1932
UALVP40747 Cast of HOLOTYPE Cast of FMNH PF 3263, original at Museum of Comparative Zoology. *Vernonaspis leonardi* **Flower and Wayland-Smith, 1952, p. 376**
USA: New York: Oneida County: about 2 mi. SE of Kenwood. Late Silurian: Salina Group, Vernon Shale. Cast of dorsal shield.

Flower and Wayland-Smith, 1952, Descr. pp. 376–378, & fig. Fig. 1, p. 377, Plate 2, Figs. 6, 7, Plate 5, Plate 15, Fig. 2. CCIS L2-245, Silurian 11, 02

Flower, R. H., and Wayland-Smith, R. 1952. Cyathaspid fishes from the Vernon Shale of New York. Bulletin of the Museum of Comparative Zoology Vol. 107(6):355–387.

Agnatha Haeckel, 1895
    Heterostraci Lankester, 1868
        Cyathaspidiformes Kiaer and Heintz, 1935
            Cyathaspidida Tarlo, 1962
            Cyathaspididae Kiaer, 1932
UALVP34185 Cast of HOLOTYPE Cast of holotype PU 17104 at Princeton University Geological Museum; UALVP40759 Cast of HOLOTYPE Cast of FMNH PF 3708, original at Princeton University, PU 17104 *Vernonaspis major* Denison, 1963, p. 110
Collectors: Bamber, E. W., and MacDonald, W. D., 1960
Canada: Yukon Territory, Beaver River. Late Silurian. Dorsal shield. Denison, 1963, Descr. pp. 110–113, & fig. Fig. 62, p. 111, Fig. 63, p. 112. UALVP40759 in CCIS L2-245, Silurian 11, 02; UALVP34185 in CCIS L2-245, Silurian 11, 06

Denison, Robert H. 1963. New Silurian Heterostraci from southeastern Yukon. Fieldiana: Geology Vol. 14(7): 105–141.

Agnatha Haeckel, 1895
    Heterostraci Lankester, 1868
        Cyathaspidiformes Kiaer and Heintz, 1935
            Cyathaspidida Tarlo, 1962
            Cyathaspididae Kiaer, 1932
UALVP40757 Cast of HOLOTYPE, Cast of FMNH PF 3679, original at National Museum of Canada, NMC 10036. *Vernonaspis sekwiae* Denison, 1964, pp. 382–384
Canada: Northwest Territories: beside Keele River, top of Mt. Sekwi, Mount Sekwi, 63°28′N. Lat., 128°40′W. Long.

Silurian-Silurian Late. Cast of incomplete dorsal shield. Denison, 1964, Descr. 382–384, & fig. Fig. 123, A-C, p. 382, Fig. 124, p. 383. CCIS L2-245, Silurian 11, 02

Denison, Robert H. 1964. The Cyathaspididae. A family of Silurian and Devonian jawless vertebrates Fieldiana: Geology Vol. 13(5):309–473.

PARATYPES:

UALVP34230 cast, ventral shield, NMC 19720, original at National Museum of Canada. CCIS L2-245, Silurian 10, 08

UALVP34238 cast of paratype, cast of NMC 19700, original at National Museum of Canada, dorsal shield, figured specimen. CCIS L2-245, Silurian 10, 08

UALVP40758 cast of paratype, cast of FMNH PF 3673, cast of crushed dorsal shield. Denison, 1964, Descr. 383–384. CCIS L2-245, Silurian 11, 02

Agnatha Haeckel, 1895
    Heterostraci Lankester, 1868
        Cyathaspidiformes Kiaer and Heintz, 1935
            Cyathaspidida Tarlo, 1962
            Cyathaspididae Kiaer, 1932

UALVP34232 cast of HOLOTYPE Cast of NMC 19713 at National Musuem of Canada *Vernonaspis tortucosta* Dineley and Loeffler, 1976, p. 62

Canada: Northwest Territories. Silurian-Silurian Late-Ludlovian? Delorme Group? Cast of dorsal shield. Dineley and Loeffler, 1976, Descr. pp. 62–63, & fig. Plate 6, Fig. 12, p. 55. CCIS L2-245, Silurian 10, 08

Dineley, D. L., and Loeffler, E. J. 1976. Ostracoderm faunas of the Delorme and associated Siluro-Devonian formations North West Territories, Canada Special Papers in Palaeontology, No. 18:1–214.

Agnatha Haeckel, 1895
    Heterostraci Lankester, 1868

Cyathaspidiformes Kiaer and Heintz, 1935
Cyathaspidida Tarlo, 1962
Cyathaspididae Kiaer, 1932
Ctenaspidinae Denison, 1964

UALVP59255 Cast of HOLOTYPE Cast of NMC 21700 at National Museum of Canada *Arctictenaspis obruchevi* (Dineley, 1976) formerly *Ctenaspis obruchevi* Dineley, 1976, p. 31

Canada: Nunavut: Prince of Wales Island, Localiy A. Devonian-Devonian Early (Gedinnian -?Siegenian); Peel Sound. Imperfect dorsal shield and tail. Dineley, 1976, Descr. p. 31, & fig. Fig. 1, A, p. 29; Fig. 2, A, p. 30. CCIS L2-245, Devonian 17, 02

Dineley, D. L. 1976. New species of *Ctenaspis* (Ostracodermi) from the Devonian of Arctic Canada. pp. 26–44. *IN:* Churcher, C. S. (editor) Athlon essays in honour of Loris Shano Russell. University of Toronto Press, Toronto, Ontario, Canada.

<u>Casts of PARATYPES:</u>

UALVP59256 Cast of of NMC 21767 Impression of scales of left flank of body and tail. Dineley, 1976, Descr. p. 31, & fig. Fig. 2, C, p. 30. CCIS L2-245, Devonian 17, 02

UALVP59257 Cast of of NMC 21768 Ventral shield, scaled body and tail. Dineley, 1976, Descr. p. 31, & fig. Fig. 2, D, p. 30. CCIS L2-245, Devonian 17, 02

Agnatha Haeckel, 1895
Heterostraci Lankester, 1868
Cyathaspidiformes Kiaer and Heintz, 1935
Cyathaspidida Tarlo, 1962
Cyathaspididae Kiaer, 1932
Tolypelepidinae Denison, 1964

UALVP34268 Cast of HOLOTYPE Cast of NMC 19767 at National Museum of Canada. *Asketaspis interstincta* Dineley and Loeffler, 1976, p. 57

Canada: Northwest Territories. Silurian-Silurian Late-Ludlovian? Delorme Group? Cast of incomplete dorsal shield. Dineley and

Loeffler, 1976, Descr. pp. 56–58, & fig. Plate 6, Fig. 1, p. 55, Text-Fig. 22, p. 57. CCIS L2-245, Silurian 10, 09

Dineley, D. L., and Loeffler, E. J. 1976. Ostracoderm faunas of the Delorme and associated Siluro-Devonian formations North West Territories, Canada Special Papers in Palaeontology. No. 18:1–214.

Additional specimen *not a PARATYPE:*
UALVP34258 cast of NMC 13220, incomplete dorsal shield, figured specimen. CCIS L2-245, Silurian 10, 09

Agnatha Haeckel, 1895
   Heterostraci Lankester, 1868
      Cyathaspidiformes Kiaer and Heintz, 1935
         Cyathaspidida Tarlo, 1962
            Cyathaspididae Kiaer, 1932
               Tolypelepidinae Denison, 1964
UALVP34248 cast of HOLOTYPE Cast of NMC 13768 *Tolypelepis lenzi* **Dineley and Loeffler, 1976, p. 53**
Canada: Northwest Territories. Silurian-Silurian Late-Ludlovian? Delorme Group? Dorsal shield, variation on dorsal shield. Dineley and Loeffler, 1976, Descr. pp. 53–56, & fig. Plate 6, Fig. 3, p. 55, Text-Fig. 21, p. 54. CCIS L2-245, Silurian 10, 09

Dineley, D. L., and Loeffler, E. J. 1976. Ostracoderm faunas of the Delorme and associated Siluro-Devonian formations North West Territories, Canada Special Papers in Palaeontology, No. 18:1–214.

Agnatha Haeckel, 1895
   Heterostraci Lankester, 1868
      Cyathaspidiformes Kiaer and Heintz, 1935
         Irregulareaspididae Denison, 1964
UALVP34187 Cast of HOLOTYPE Cast of HOLOTYPE PU 17088 at Princeton University Geological Museum *Dikenaspis yukonensis* **Denison, 1963, pp. 116–129**
Collectors: Bamber, E. W., and MacDonald, W. D., 1959.

Canada: SE Yukon Territory, Beaver River, 60°27′N. Lat., 125°46′W. Long. Silurian-Silurian Late-Ludlovian? Delorme Group? Dorsal shield, and posterior part of ventral shield. Denison, 1963, Descr. pp. 116–129, & fig. Fig. 66 (p. 118), Fig. 67 (p. 119). CCIS L2-245, Silurian 11, 02; also CCIS L2-245, Silurian 11, 06

Denison, Robert H. 1963. New Silurian Heterostraci from southeastern Yukon. Fieldiana: Geology, 14(7):105–141.

Agnatha Haeckel, 1895
    Heterostraci Lankester, 1868
        Cyathaspidiformes Kiaer and Heintz, 1935
            Irregulareaspididae Denison, 1964
UALVP34266 Cast of HOLOTYPE Cast of NMC 19673. *Nahanniaspis mackenziei* **Dineley and Loeffler, 1976, p. 92**
Canada: Northwest Territories. Silurian-Silurian Late-Ludlovian? Delorme Group? An almost complete articulated specimen. Dineley and Loeffler, 1976, Descr. pp. 92–101, & fig. Plate 14, p. 97, Text-Fig. 35, A-F, p. 99, Text-Fig. 36, p. 100. CCIS L2-245, Silurian 10, 09

Dineley, D. L., and Loeffler, E. J. 1976. Ostracoderm faunas of the Delorme and associated Siluro-Devonian formations North West Territories, Canada. Special Papers in Palaeontology, No. 18:1–214.

PARATYPES:
UALVP34263 dorsal shield Cast of NMC 13212. Dineley and Loeffler, 1976, Descr. pp. 92–101, & fig. Plate 13, Fig. 1, p. 95, Text-Fig. 31, p. 93. CCIS L2-245, Silurian 10, 09
UALVP34265 cast, NMC 19659b, lateral line canal system on dorsal shield. Dineley and Loeffler, 1976, Descr. p. 92, & fig. Plate 13, Fig. 6, p. 95, Text-Fig. 32, A, p. 94. Listed as "other material" but figured in original description, so it is a paratype. CCIS L2-245, Silurian 10, 09

Agnatha Haeckel, 1895
    Heterostraci Lankester, 1868

Cyathaspidiformes Kiaer and Heintz, 1935
Cyathaspidida Tarlo, 1962
Poraspididae Kiaer, 1932
UALVP40754 Cast of HOLOTYPE Cast of *Palaeaspis ameri-cana* Claypole, 1885, Cast of FMNH PF 3822, original at Los Angeles County Museum, LACM 6393. *Americaspis americana* (Claypole, 1885, p. 48)
The United States: Pennsylvania: Perry County. Late Silurian: Wills Creek Fm.; Landisburg sandstone member. Claypole, 1885, Descr. pp. 48–64, & fig. Fig. 7. CCIS L2-245, Silurian 11, 02

Claypole, E. W. 1885. On the recent discovery of pteraspidian fish in the Upper Silurian rocks of North America. Quarterly Journal of the Geological Society of London Vol. 41:48–64.

Agnatha Haeckel, 1895
  Heterostraci Lankester, 1868
    Cyathaspidiformes Kiaer and Heintz, 1935
      Cyathaspidida Tarlo, 1962
        Poraspididae Kiaer, 1932
UALVP55739 (formerly UALVP49532) *Poraspis sp. nov.* A, Hawthorn, 2009.
[UALVP49532]. Glaser seems to have overwritten this catalogue entry which was originally the holotype specimen for a new species in Hawthorn, 2009, unpublished M.Sc. thesis. See new catalogue entry UALVP55739.]
Canada: Northwest Territories, Mackenzie Mountains, MOTH: Man on the Hill #1, Devonian-Devonian Early-Lochkovian, Road River Fm. Dorsal & ventral shield, dorsal median scale, left dorsolateral scale, left ventrolateral scale, 1 indeterminate trunk scale. Hawthorn, 2009, Descr. pp. 69–72, p. 77, Tables 2.2 p. 34, 2.3 p. 35, & fig. Figs. 2.8, A, B, p. 70 CCIS L2-245, Devonian 03, 16

Hawthorn, Jessica Rae 2009. New Poraspidine Heterostracans from the Lochkovian (Early Devonian) Man on the Hill local-ity, Mackenzie Mountains, Northwest Territories, Canada, and the phylogeny and evolutionary history of Poraspidinae. M.Sc. thesis. Department of Biological Sciences, University of Alberta, Edmonton, Alberta, Canada. 156 pp.

Agnatha Haeckel, 1895
 Heterostraci Lankester, 1868
  Cyathaspidiformes Kiaer and Heintz, 1935
   Cyathaspidida Tarlo, 1962
    Poraspididae Kiaer, 1932
UALVP32886 *Poraspis sp. nov.* **B, Hawthorn, 2009.**
Collector: Soehn, Kenneth L., 1990
Canada: Northwest Territories, Mackenzie Mountains, MOTH: Man on the Hill #1, Devonian-Devonian Early-Lochkovian, Road River Fm. Plate, Dorsal shield. Hawthorn, 2009, Descr. pp. 72–74, Table 2.2, p. 34, Table 2.3, p. 35, & fig. Fig. 2.9A, p. 73. CCIS L2-245, Devonian 03, 16

 Hawthorn, Jessica Rae 2009. New Poraspidine Heterostracans from the Lochkovian (Early Devonian) Man on the Hill locality, Mackenzie Mountains, Northwest Territories, Canada, and the phylogeny and evolutionary history of Poraspidinae. M.Sc. thesis. Department of Biological Sciences, University of Alberta, Edmonton, Alberta, Canada. 156 pp.
PARATYPES:
UALVP45941 Cephalothorax Dorsal shield. Hawthorn, 2009, Descr. pp. 72–74, Table 2.2, p. 34, Table 2.3, p. 35, & fig. Fig. 2.9B, p. 73 CCIS L2-245, Devonian 03, 16
UALVP47062 Dorsal shield anterior right part. Hawthorn, 2009, Descr. pp. 72–74, Table 2.2 p. 34, Table 2.3 p. 35, & fig. Fig. 2.9, C, p. 73 CCIS L2-245, Devonian 03, 16

Agnatha Haeckel, 1895
 Heterostraci Lankester, 1868
  Cyathaspidiformes Kiaer and Heintz, 1935
   Cyathaspidida Tarlo, 1962
    Poraspididae Kiaer, 1932

UALVP43232 *Poraspis sp. nov.* **C, Hawthorn, 2009**
Collector: Wilson, Mark V. H., 1998.
Northwest Territories, Mackenzie Mountains, MOTH: Man on the Hill #1, Devonian-Devonian Early-Lochkovian, Delorme Group. 1 piece, Dorsal shield. Hawthorn, 2009, Descr. pp. 75–77, Table 2.2, p. 34, Table 2.3, p. 35, & fig. Fig. 2.10, p. 76 CCIS L2-245, Devonian 03, 16

Hawthorn, Jessica Rae 2009. New Poraspidine Heterostracans from the Lochkovian (Early Devonian) Man on the Hill locality, Mackenzie Mountains, Northwest Territories, Canada, and the phylogeny and evolutionary history of Poraspidinae. M.Sc. thesis. Department of Biological Sciences, University of Alberta, Edmonton, Alberta, Canada. 156 pp.

Agnatha Haeckel, 1895
 Heterostraci Lankester, 1868
  Cyathaspidiformes Kiaer and Heintz, 1935
   Cyathaspidida Tarlo, 1962
    Poraspididae Kiaer, 1932
UALVP41246 *Poraspis sp. nov.* **D, Hawthorn, 2009**
Collector: Hanke, Gavin, 1998.
Canada: Northwest Territories, Mackenzie Mountains, MOTH: Man on the Hill #1, Devonian-Devonian Early-Lochkovian, Road River Fm. Dorsal shield. Hawthorn, 2009, Descr. pp. 77–80, Table 2.2, p. 34, Table 2.3 p. 35, & fig. Fig. 2.11, p. 78

Hawthorn, Jessica Rae 2009. New Poraspidine Heterostracans from the Lochkovian (Early Devonian) Man on the Hill locality, Mackenzie Mountains, Northwest Territories, Canada, and the phylogeny and evolutionary history of Poraspidinae. M.Sc. thesis. Department of Biological Sciences, University of Alberta, Edmonton, Alberta, Canada. 156 pp.

Agnatha Haeckel, 1895
 Heterostraci Lankester, 1868
  Cyathaspidiformes Kiaer and Heintz, 1935

Cyathaspidida Tarlo, 1962
Poraspididae Kiaer, 1932
UALVP40727 Cast of HOLOTYPE, Oslo Paleontological Museum, PMOD 203. *Poraspis magna* Kiaer, 1932, p. 14
Spitsbergen: Red Bay: Ben Nevis.
Early Devonian: Red Bay Series, Ben Nevis Group, Horizon A. Dorsal shield. Kiaer, 1932, Descr. p. 14. CCIS L2-245, Devonian 08, 12

Kiaer, Johan. 1932. The Downtonian and Devonian vertebrates of Spitsbergen. IV. Suborder Cyathaspida. Skrifter om Svalbard og Ishavet. 52:1–26.

Agnatha Haeckel, 1895
Heterostraci Lankester, 1868
Eriptychiformes Tarlo, 1962
Lepidaspididae Halstead, 1993
UALVP23138 HOLOTYPE *Lepidaspis loefflerae* (Greeniaus, 2004, p. 75 [Master of Science thesis, not published])
Canada: Northwest Territories, Mackenzie Mountains, MOTH: Man on the Hill #1, Devonian-Devonian Early-Lochkovian, Delorme Group. Posterior part of dorsal cephalothorax and tail. Greeniaus, 2004, Descr. p. 75, p. 76, p. 82, p. 83, & fig. Fig. 3.3 A–E, p. 78. Earth Sciences Museum, ESB B-01

Greeniaus, Jeffrey W. 2004. Description of Devonian tessellate heteostracans from the Northwest Territories, Canada and the growth of *Lepidaspis* Master of Science thesis University of Alberta, Edmonton, Alberta, Canada, pp. 1–119.

PARATYPES:
UALVP23172 partial (plates). Greeniaus, 2004, Descr. p. 75, & fig. Fig. 3.5, B, p. 86. CCIS L2-245, Devonian 04, 03
UALVP23280 partial specimen with scales. Greeniaus, 2004, Descr. p. 75, & fig. Fig. 3.5, A, p. 86. CCIS L2-245, Devonian 08, 02; Z-425, Unit 10, Drawer 4.
UALVP32591 fish scales; Greeniaus, 2004, p. 86. CCIS L2-245, Devonian 04, 03

UALVP43417 (formerly UALVP47001) fish scales; Greeniaus, 2004, Descr. p. 72, & fig. Fig. 3.7, A–E, p. 94. CCIS L2-245, Devonian 08, 02; **UALVP43419 (formerly UALVP47009)** This specimen was catalogued by Jeffrey Greeniaus as UALVP47009 but Raoul Mutter published this number first, changed to UALVP43419. Greeniaus, 2004, Cephalothorax and fused lateral margin in poorly articulated specimen. Greeniaus, 2004, Descr. p. 75, p. 83; & fig. Fig. 3.4, B, p. 80. CCIS L2-245, Devonian 08, 02

UALVP43420 (formerly UALVP47010) This specimen was catalogued by Jeffrey Greeniaus as UALVP47010 but Raoul Mutter published this number first, changed to UALVP43420. Thin sections A and B through cephalothorax. Greeniaus, 2004, Descr. p. 75, p. 83, & fig. Fig. 3.5, D, p. 86, Fig. 3.6, A–D, p. 88. CCIS L2-245, Devonian 04, 03

UALVP43847 partial (plates). Greeniaus, 2004, Descr. p. 75, p. 76, & fig. Fig. 3.4, A, p. 80. CCIS L2-245, Devonian 04, 03

Agnatha Haeckel, 1895
    Heterostraci Lankester, 1868
        Eriptychiformes Tarlo, 1962
            Lepidaspididae Halstead, 1993
***Lepidaspis serrata* Dineley and Loeffler, 1976**
Canada: Northwest Territories. Silurian-Silurian Late-Ludlovian? Delorme Group?

    Dineley, D. L., and Loeffler, E. J. 1976. Ostracoderm faunas of the Delorme and associated Siluro-Devonian formations North West Territories, Canada Special Papers in Palaeontology, No. 18:1–214.

PARATYPE:
UALVP34194 cast of NMC 19882 cephlothorax with lateral lamina and isolated tail scales. Dineley and Loeffler, 1976, Descr. p. 176, p. 180, & fig. Plate 29. p. 179. CCIS L2-245, Silurian 10, 08

Agnatha Haeckel, 1895
    Heterostraci Lankester, 1868

Eriptychiformes Tarlo, 1962
Tesseraspididae Berg, 1940
UALVP45700 HOLOTYPE *Grandipiscis exscopulus* **Greeniaus, 2004, p. 17 [Master of Science thesis, not published]**
Canada Northwest Territories, Mackenzie Mountains, MOTH: Man on the Hill #1, 62°32′N. Lat., 127°43′W. Long. Devonian-Devonian Early-D1-Lochkovian (Gedinnian-Siegenian), Delorme Group, Complete anterior dorsal cephalothorax. Greeniaus, 2004, Descr. p. 17, p. 18, p. 23, p. 26, & fig. Fig. 2.2, p. 20, Fig. 2.4, A, p. 25. CCIS L2-245, Devonian 03, 09

Greeniaus, Jeffrey W. 2004. Description of Devonian tessellate heteostracans from the Northwest Territories, Canada and the growth of *Lepidaspis* Master of Science thesis University of Alberta, Edmonton, Alberta, Canada, pp. 1–119.

<u>**PARATYPES:**</u>

UALVP23190 ventral shield. Greeniaus, 2004, Descr. p. 17, p. 26, & fig. Fig. 2.5, B, p. 28. CCIS L2-245, Devonian 04, 02

UALVP32523 interior of dorsal shield. Greeniaus, 2004, Descr. p. 17, & fig. Fig. 2.6, B, p. 30. CCIS L2-245, Devonian 04, 02

UALVP43411 (formerly UALVP47001) near complete cephalothorax. Greeniaus, 2004, Descr. p. 17, p. 18, p. 23, p. 26, p. 33, p. 35, & fig. Fig. 2.3, p. 22; Fig. 2.4, B, p. 25; Fig. 2.7B, p. 32. This specimen was catalogued by Jeffrey Greeniaus as UALVP47001, but Raoul Mutter published this number first, changed to UALVP43411. CCIS L2-245, Devonian 04, 02

UALVP43412 (formerly UALVP47002) interior of dorsal shield & exterior of ventral. Greeniaus, 2004, Descr. p. 17, p. 35, p. 36, & fig. Fig. 2.5A, p. 28. This specimen was catalogued by Jeffrey Greeniaus as UALVP47002 but Raoul Mutter published this number first, changed to UALVP43412. CCIS L2-245, Devonian 04, 02

UALVP43413 (formerly UALVP47003) interior of dorsal shield & exterior of ventral. Greeniaus, 2004, Descr. p. 17, p. 26, p. 33, p. 34, p. 36, & fig. Fig. 2.6, C, p. 30, Fig. 2.7A, p. 32. This specimen was catalogued by Jeffrey Greeniaus as UALVP47003, but Raoul

Mutter published this number first, changed to UALVP43413. CCIS L2-245, Devonian 04, 02

UALVP43414 (formerly UALVP47004) posterior of ventral shield & anterior part of tail region. Greeniaus, 2004, Descr. (AS UALVP47004, NOW UALVP43414), Descr. p. 26, p. 33, p. 34, & fig. Fig. 2.6, A, p. 30. This specimen was catalogued by Jeffrey Greeniaus as UALVP47004, but Raoul Mutter published this number first, changed to UALVP43414. CCIS L2-245, Devonian 04, 02

Agnatha Haeckel, 1895
    Heterostraci Lankester, 1868
        Eriptychiformes Tarlo, 1962
            Tesseraspididae Berg, 1940
UALVP43415 HOLOTYPE *Quasimodaspis canadensis* Greeniaus, 2004 [Master of Science thesis, not published] Canada: Northwest Territories, Mackenzie Mountains, MOTH: Man on the Hill #1, Devonian-Devonian Early-Lochkovian, Delorme Group. Dorsal portion of cephalothorax, median plate. Greeniaus, 2004, Descr. p. 52, p. 53, p. 54. This specimen was catalogued by Jeffrey Greeniaus as UALVP47005 but Raoul Mutter published this number first, changed to UALVP43415. CCIS L2-245, Devonian 04, 04

    Greeniaus, Jeffrey W. 2004. Description of Devonian tessellate heteostracans from the Northwest Territories, Canada, and the growth of *Lepidaspis* Master of Science thesis University of Alberta, Edmonton, Alberta, Canada, pp. 1–119.

PARATYPES:

UALVP32495 both portions of the median plate of the dorsal cephalothorax. Greeniaus, 2004, Descr. p. 52 CCIS L2-245, Devonian 04, 04

UALVP43416 median plate of dorsal cephalothorax. Greeniaus, 2004, Descr. p. 52, p. 53, p. 54, p. 55, & fig. Fig. 2.10, B, p. 50. This specimen was catalogued by Jeffrey Greeniaus as UALVP47006, but Raoul Mutter published this number first, changed to UALVP43416. CCIS L2-245, Devonian 04, 04

Agnatha Haeckel, 1895
   Heterostraci Lankester, 1868
      Eriptychiformes Tarlo, 1962
         Tesseraspididae Berg, 1940
UALVP43845 HOLOTYPE *Quasimodaspis mackenziensis* **Greeniaus, 2004. P. 47. [Master of Science thesis, not published]**
Canada: Northwest Territories, Mackenzie Mountains, MOTH: Man on the Hill #1, Devonian-Devonian Early-Lochkovian, Delorme Group. Partial left side of cephalothorax. Greeniaus, 2004, Descr. p. 47, & fig. Fig. 2.10, A, p. 50. CCIS L2-245, Devonian 04, 04

    Greeniaus, Jeffrey W. 2004. Description of Devonian tessellate heteostracans from the Northwest Territories, Canada and the growth of *Lepidaspis* Master of Science thesis University of Alberta, Edmonton, Alberta, Canada, pp. 1–119.

Agnatha Haeckel, 1895
   Heterostraci Lankester, 1868
      Pteraspidiformes Berg, 1940
         Pteraspididae Claypole, 1885
UALVP34195 cast of HOLOTYPE Cast of NMC 19993. *Canadapteraspis alocostomata* **Dineley and Loeffler, 1976, p. 113**
Canada: Northwest Territories. Silurian-Silurian Late-Ludlovian? Delorme Group? Cast of incomplete dorsal shield. Dineley and Loeffler, 1976, Descr. pp. 113–117, & fig. Plate 17, Fig. 3, p. 125. CCIS L2-245, Silurian 10, 08

    Dineley, D. L., and Loeffler, E. J. 1976. Ostracoderm faunas of the Delorme and associated Siluro-Devonian formations North West Territories, Canada. Special Papers in Palaeontology, No. 18:1–214.

PARATYPE:
UALVP34231 cast, ventral disc, Cast of NMC 21458. Dineley and Loeffler, 1976, Descr. pp. 113–117, & fig. Plate 15, Fig. 2, p. 115, Text-Fig. 43, B, p. 114. CCIS L2-245, Silurian 10, 08

Agnatha Haeckel, 1895
  Heterostraci Lankester, 1868
    Phialaspidiformes Berg, 1935
      Traquairaspididae Kiaer, 1932
UALVP34234 cast of HOLOTYPE Cast of NMC 21402?
Traquairaspis adunata **Dineley and Loeffler, 1976, p. 20**
Canada: Northwest Territories. Silurian-Silurian Late-Ludlovian?
Delorme Group? Variation in ornamentaion of dorsal shield.
Dineley and Loeffler, 1976, Descr. pp. 20–22, & fig. Plate 1, Fig. 5,
p. 21, Text-Fig. 3, a, b, p. 20. CCIS L2-245, UALVP34234 in CCIS
L2-245, Silurian 10, 08

Dineley, D. L., and Loeffler, E. J. 1976. Ostracoderm faunas
of the Delorme and associated Siluro-Devonian formations North
West Territories, Canada Special Papers in Palaeontology, No.
18:1–214.

Agnatha Haeckel, 1895
  Heterostraci Lankester, 1868
    Phialaspidiformes Berg, 1935
      Traquairaspididae Kiaer, 1932
**UALVP34191 Cast of HOLOTYPE Cast of HOLOTYPE PU
17388 at Princeton University Geological Museum** *Traquairaspis*
(*Rimasventeraspis***) *angusta* Denison, 1963, pp. 133–135**
Collectors: Bamber, E. W., and MacDonald, W. D., 1959.
Canada: Southeastern Yukon Territory, Beaver River, 60°27″N.
Lat., 125°47.5′W. Long. Silurian-Silurian Late-Ludlovian? Delorme
Group? Ventral disc. Denison, 1963, Descr. pp. 133–135, & fig.
Fig. 78, p. 133, Fig. 79, p. 135. CCIS L2-245, Silurian 11, 06

Denison, Robert H. 1963. New Silurian Heterostraci from
southeastern Yukon. Fieldiana: Geology, 14(7):105–141.

Agnatha Haeckel, 1895
  Heterostraci Lankester, 1868
    Phialaspidiformes Berg, 1935
      Traquairaspididae Kiaer, 1932

UALVP34198 cast of HOLOTYPE Cast of NMC 21418?
Traquairaspis broadi **Dineley and Loeffler, 1976, p. 29**
Canada: Northwest Territories. Silurian-Silurian Late-Ludlovian, Delorme Group? Cast of large ventral shield, lacking a ventral primordium. Dineley and Loeffler, 1976, Descr. pp. 29–30, & fig. Plate 2, Fig. 3, p. 27, Text-Fig. 8, p. 29. CCIS L2-245, Silurian 10, 08

Dineley, D. L., and Loeffler, E. J. 1976. Ostracoderm faunas of the Delorme and associated Siluro-Devonian formations North West Territories, Canada. Special Papers in Palaeontology, No. 18:1–214.

Agnatha Haeckel, 1895
Heterostraci Lankester, 1868
Phialaspidiformes Berg, 1935
Traquairaspididae Kiaer, 1932
UALVP34242 Cast of HOLOTYPE Cast of NMC 21411?
Traquairaspis guttata **Dineley and Loeffler, 1976, p. 22**
Canada: Northwest Territories. Silurian-Silurian Late-Ludlovian? Delorme Group? An incomplete ventral shield. Dineley and Loeffler, 1976, Descr. pp. 22–24, & fig. Plate 1, Fig. 4, p. 21, Text-Fig. 4, p. 23. CCIS L2-245, Silurian 10, 08

Dineley, D. L., and Loeffler, E. J. 1976. Ostracoderm faunas of the Delorme and associated Siluro-Devonian formations North West Territories, Canada Special Papers in Palaeontology, No. 18:1–214.

Agnatha Haeckel, 1895
Heterostraci Lankester, 1868
Phialaspidiformes Berg, 1935
Traquairaspididae Kiaer, 1932
UALVP34197 Cast of HOLOTYPE Cast of NMC 21436?
Traquairaspis lemniscata **Dineley and Loeffler, 1976, p. 30**
Canada: Northwest Territories. Silurian-Silurian Late-Ludlovian? Delorme Group? Cast of incomplete ventral shield Dineley and Loeffler, 1976, Descr. pp. 30–31, & fig. Plate 4, Fig. *2, p. 36, Text-Fig. 9, p. 31. CCIS L2-245, Silurian 10, 08

*Note: According to figure information for Fig. 2 on p. 36, the figure is of NMC 21435, but the information on p. 30 for the holotype, Fig. 2 is of NMC 21436.

Dineley, D. L., and Loeffler, E. J. 1976. Ostracoderm faunas of the Delorme and associated Siluro-Devonian formations North West Territories, Canada Special Papers in Palaeontology, No. 18:1–214.

Agnatha Haeckel, 1895
    Heterostraci Lankester, 1868
        Phialaspidiformes Berg, 1935
            Traquairaspididae Kiaer, 1932
UALVP34235 cast of HOLOTYPE Cast of NMC 19782?
Traquairaspis mackenziensis **Dineley and Loeffler, 1976, p. 31**
Canada: Northwest Territories. Silurian-Silurian Late-Ludlovian? Delorme Group? Cast, incomplete dorsal shield, variation in ornamentation on anterior part of dorsal shield. Dineley and Loeffler, 1976, Descr. pp. 31–38, & fig. Plate 3, Fig. 1, p. 33, Text-Fig. 12, p. 34. CCIS L2-245, Silurian 10, 08

Dineley, D. L., and Loeffler, E. J. 1976. Ostracoderm faunas of the Delorme and associated Siluro-Devonian formations North West Territories, Canada. Special Papers in Palaeontology, No. 18:1–214.
PARATYPES:
UALVP34233 cast of NMC l9791, relationship of lateral line pores 2, in ornamentation on lateral margins of dorsal shield. Dineley and Loeffler, 1976, Descr. pp. 35, & fig. Plate 3, Fig. 3, p. 33; Text Fig. 10, p. 32. CCIS L2-245, Silurian 10, 08
UALVP34236 cast, incomplete dorsal shield, NMC 19790. Dineley and Loeffler, 1976, Descr. pp. 31, 36, & fig. Plate 3, Fig. 4, p. 33; Text Fig. 11, p. 34. CCIS L2-245, Silurian 10, 08
UALVP34237 cast of NMC 19801, incomplete ventral shield, relationship of lateral line pores to ornamentation of ventral shield. Dineley and Loeffler, 1976, Descr. pp. 31, 36, & fig. Plate 3, Fig. 2, p. 33; Text Fig. 14, p. 34. CCIS L2-245, Silurian 10, 08

Agnatha Haeckel, 1895
 Heterostraci Lankester, 1868
  Phialaspidiformes Berg, 1935
   Traquairaspididae Kiaer, 1932
UALVP34267 cast of HOLOTYPE Cast of NMC 21395 at
National Museum of Canada ?*Traquairaspis poolei* Dineley
and Loeffler, 1976, p. 18
Canada: Northwest Territories. Silurian-Silurian Late-Ludlovian?
Delorme Group? Cast of incomplete dorsal shield Dineley and
Loeffler, 1976, Descr. pp. 18–19, & fig. Plate 1, Fig. 3, p. 21, Text-
Fig. 2, p. 19. CCIS L2-245, Silurian 10, 09
    Dineley, D. L., and Loeffler, E. J. 1976. Ostracoderm faunas
of the Delorme and associated Siluro-Devonian formations North
West Territories, Canada Special Papers in Palaeontology, No.
18:1–214.

Agnatha Haeckel, 1895
 Heterostraci Lankester, 1868
  Phialaspidiformes Berg, 1935
   Traquairaspididae Kiaer, 1932
UALVP34199 cast of HOLOTYPE cast of NMC 21416
?*Traquairaspis pustulata* Dineley and Loeffler, 1976, p. 24
Canada: Northwest Territories. Silurian-Silurian Late-Ludlovian?
Delorme Group? Incomplete ventral shield. Dineley and Loeffler,
1976, Descr. pp. 24–28, & fig. Plate 2, Fig. 1, p. 27, text fig. 6, a, b, c,
p. 26. CCIS L2-245, Silurian 10, 08
    Dineley, D. L., and Loeffler, E. J. 1976. Ostracoderm faunas
of the Delorme and associated Siluro-Devonian formations North
West Territories, Canada Special Papers in Palaeontology, No.
18:1–214.
PARATYPES:
UALVP34196 cast of NMC 21447 Incomplete ventral shield.
Dineley and Loeffler, 1976, Descr. p. 22 & fig. Plate 2, Fig. 4, p. 27.
CCIS L2-245, Silurian 10, 08

Agnatha Haeckel, 1895
    Heterostraci Lankester, 1868
        Phialaspidiformes Berg, 1935
            Traquairaspididae Kiaer, 1932
UALVP34247 cast of HOLOTYPE Cast of NMC 21422
?*Traquairaspis retusa* Dineley and Loeffler, 1976, p. 24
Canada: Northwest Territories. Silurian-Silurian Late-Ludlovian? Delorme Group? Cast of incomplete dorsal disc Dineley and Loeffler, 1976, Descr. p. 24 & fig. Fig. Plate 1, Fig. 1, p. 21, Text-Fig. 5, p. 25. CCIS L2-245, Silurian 10, 09

    Dineley, D. L., and Loeffler, E. J. 1976. Ostracoderm faunas of the Delorme and associated Siluro-Devonian formations North West Territories, Canada. Special Papers in Palaeontology, No. 18:1–214.

Agnatha Haeckel, 1895
    Osteostraci Lankester, 1868
        Cornuata Janvier, 1985
            Zenaspidida Stensiö, 1958
                Superciliaspididae Scott and Wilson, 2014
UALVP32408 PARATYPE *Dentapelta loefflerae* Scott and Wilson, 2014
Collector: Lindoe, L. Allan, 1990 Canada: Northwest Territories, Mackenzie Mountains, MOTH: Man on the Hill #1, Devonian-Devonian Early-Lochkovian, Road River Fm. Fish. Scott and Wilson, 2014, Descr. p. 10, & Fig. 4, C, p. 11. CCIS L2-245, Devonian 03, 03

    Scott, Bradley R. and Wilson, Mark V. H. 2014. The Superciliaspididae, a new family of Early Devonian Osteostraci (jawless vertebrates) from northern Canada, with two new genera and three new species. Journal of Systematic Palaeontology Vol. 13(3):167–187. [pp. 1–21. Published online: 13 Mar 2014.]
<u>PARATYPE:</u>
UALVP43640 partial dorsal head shield, with median spine and posterior pointed denticulation on teseerae. One piece. Scott and

Wilson, 2014, Descr. p. 2, Table 1, p. 7, p. 10, Table 2, p. 13, & fig. Fig. 4, B, p. 11, Fig. 5, A-C, p 12. CCIS L2-245, Devonian 03, 08

Agnatha Haeckel, 1895
 Osteostraci Lankester, 1868
  Cornuata Janvier, 1985
   Zenaspidida Stensiö, 1958
    Superciliaspididae Scott and Wilson, 2014
UALVP43639 HOLOTYPE *Glabrapelta cristata* Scott and Wilson, 2014, p. 4
Canada: Northwest Territories, Mackenzie Mountains, MOTH: Man on the Hill #2, Devonian-Devonian Early-Lochkovian, Road River Fm. Nearly complete head shield, dorsal view. Scott and Wilson, 2014, Descr. p. 4, Table 1, p. 7, Table 2, p. 13, & fig. Fig. 2, A, B, p. 5, Fig. 7, B, p.15, CCIS L2-245, Devonian 03, 08
 Scott, Bradley R. and Wilson, Mark V. H. 2014. The Superciliaspididae, a new family of Early Devonian Osteostraci (jawless vertebrates) from northern Canada, with two new genera and three new species. Journal of Systematic Palaeontology Vol. 13(3):167–187. [pp. 1–21. Published online: 13 Mar 2014.]
PARATYPES:
UALVP41731 This specimen is not in list of paratypes but is cited in text. Scott and Wilson, 2014, Descr. Table 1, p. 7, p. 8. CCIS L2-245, Devonian 03, 06
UALVP42731 (now UALVP55517) a partial head shield including regions anterior to the median dorsal crest Scott and Wilson, 2014, Descr. Table 1, p. 7, p. 8 [Double use of catalogue number: *Glabrapelta cristata* Scott and Wilson, 2014 paratype & figured specimen is now UALVP55517] CCIS L2-245, Devonian 03, 06
UALVP43214 a partial head shield with some depth, Scott and Wilson, 2014, fig. Fig. 2D, p. 5. CCIS L2-245, Devonian 03, 08
UALVP55517 (was UALVP42731) partial head shield including regions anterior to the median dorsal crest. Scott and Wilson, 2014, Descr. p. 4, & fig. Fig. 2, C. p. 5. [Double use of catalogue number UALVP42731 *Microcosmodon conus* Isolated Rm1 Mammalia]

Agnatha Haeckel, 1895
   Osteostraci Lankester, 1868
      Cornuata Janvier, 1985
         Zenaspidida Stensiö, 1958
            Superciliaspididae Scott and Wilson, 2014
UALVP55518 HOLOTYPE Double use of catalogue number UALVP43083, recatalogued as UALVP55518 *Glabrapelta minima* **Scott and Wilson, 2014, p. 8**
Canada: Northwest Territories, Mackenzie Mountains, MOTH: Man on the Hill #1, Devonian-Devonian Early-D1-Lochkovian (Gedinnian-Siegenian), Road Fm. Right side of small head shield in dorsal view. Scott and Wilson, 2014, Descr. p. 8-10, Table 1, p. 7, Table 2, p. 13, & fig. Fig. 3A–C, p. 6, Fig. 7A, p. 15, [Double use of catalogue number UALVP43083 Isolated Rm3 Zack *et al.*, 2005 Table 1, p. 813, p. 816 *Gingerichia hystrix* Zack *et al.*, 2005 PARATYPE].
    Scott, Bradley R. and Wilson, Mark V. H. 2014. The Superciliaspididae, a new family of Early Devonian Osteostraci (jawless vertebrates) from northern Canada, with two new genera and three new species Journal of Systematic Palaeontology Vol. 13(3):167–187. [pp. 1–21. Published online: 13 Mar 2014.]

Agnatha Haeckel, 1895
   Osteostraci Lankester, 1868
      Cornuata Janvier, 1981
         Zenaspidida Stensiö, 1958
            Zenaspididae Stensiö, 1958
UALVP32970 Holotype *Diademaspis? mackenziensis* **Adrain and Wilson, 1994, p. 315**
Collector: Soehn, Kenneth L., 1990
Canada Northwest Territories, Mackenzie Mountains, MOTH: Man on the Hill #1, Devonian-Devonian Early-Lochkovian, Road River Fm., a partial cephalic shield with the medial area well preserved and a few trunk scales attached. Adrain and Wilson, 1994, Descr. p. 315, & fig. Fig. 10A, p. 315

Adrain, Jonathan M., and Wilson, Mark V. H. 1994. Early Devonian cephalaspids (Vertebrata: Osteostraci: Cornuata) from the southern Mackenzie Mountains, N.W.T., Canada. Journal of Vertebrate Paleontology, 14(3):301–319.

PARATYPE:

UALVP32971 an intact, largely uncompressed head with inner surface of dorsal shield exposed. Adrain and Wilson, 1994, fig. Fig. 11, p. 316. CCIS L2-245, Devonian 03, 04

Agnatha Haeckel, 1895
  Osteostraci Lankester, 1868
    Cornuata Janvier, 1981
      Zenaspidida Stensiö, 1958
        Zenaspididae Stensiö, 1958
UALVP53224 HOLOTYPE *Machairaspis serrata* **Scott and Wilson, 2013, pp. 128–132**
Canada: Northwest Territories, Mackenzie Mountains, MOTH: Man on the Hill #1, Devonian-Devonian Early-D1-Lochkovian (Gedinnian-Siegenian), Delorme Group, head shield in lateral view & associated lateral scale, 5 pieces. Scott and Wilson, 2013, Descr. pp. 128–132, & fig.Fig. 1, p. 128, Fig. 2, A-C, p. 130, Fig. 3, A, p. 131, Fig. 4, p. 131 CCIS L2-245, Devonian 03, 09

Scott, Bradley R. and Wilson, Mark V. H. 2013. A new species of osteostracan from the Lochkovian (Early Devonian) of the Mackenzie Mountains, with comments on body size, growth, and geographic distribution in the genus *Machairaspis*. Canadian Journal of Earth Sciences Vol. 50(2): 127–134

PARATYPES

UALVP23154 a very small dorsal cephalic spine in left lateral view shield fragment; paratype for *Waengsjoeaspis nahanniensis* Adrain and Wilson, 1994, but reidentified as *Waengsjoeaspis* sp. in Scott and Wilson, 2014, p. 1246. Wilson and Caldwell, 1998 J. Vert. Paleo. Vol. 18(1):10–29, Descr. p. 20–21 paratype for *Furcacauda fredhomae*. Paratype for *Machairaspis serrata* Scott and Wilson, 2013, Fig. 3, B, p. 131; *Polymerolepis whitei* according to Hanke *et al.* (2013); CCIS L2-245, Devonian 03, 02

UALVP32435 incomplete dorsal cephalic spine in right lateral view. Scott and Wilson, 2013, Descr. p. 128, p. 132 & fig. Fig. 3, C, p. 131

Agnatha Haeckel, 1895
    Osteostraci Lankester, 1868
        Cornuata Janvier, 1985
            Family *incertae sedis*
UALVP19140 HOLOTYPE *Waengsjoeaspis nahanniensis* **Adrain and Wilson, 1994**
Collector: Chatterton, Brian D. E. 1983 Canada: Northwest Territories, Mackenzie Mountains, MOTH: Man on the Hill #1, Devonian-Devonian Early-Lochkovian, Road River Fm., 1/2 of dorsal head shield, a partial (left side) cephalic shield and cornu, with orbit, dorsal sensory field, pineal, and nasohypophyseal region well preserved. Adrain and Wilson, 1994, Fig. 7, A, p. 311; Figs. 8, A, 8, C, p. 313, non 7, B, 8, B (original description). Biological Sciences Building Z-425 Unit 10, Drawer 5.

    Adrain, Jonathan M., and Wilson, Mark V. H. 1994. Early Devonian cephalaspids (Vertebrata: Osteostraci: Cornuata) from the southern Mackenzie Mountains, N.W.T., Canada. Journal of Vertebrate Paleontology, 14(3):301–319.

<u>PARATYPES:</u>

UALVP19141 right half of margin of head shield, orbits, and nasohypophyseal region; Adrain and Wilson, 1994, Descr, p. 312 CCIS L2-245, Devonian 03, 02

UALVP19145 part of head shield, with intact left "pectoral" fin and articulated trunk scales. Adrain and Wilson, 1994, Fig. 7, B, p. 311. Adrain and Wilson, 1994, but reidentified as *Waengsjoeaspis* sp. in Scott and Wilson, 2012 Earth Sciences Museum, ESB B-01

UALVP19147 part of dorsal shield. Adrain and Wilson, 1994, but reidentified as *Waengsjoeaspis* sp. in Scott and Wilson, 2012. CCIS L2-245, Devonian 5, 3

UALVP19148 part of head shield. Adrain and Wilson, 1994, p. 312, but reidentified as *Waengsjoeaspis* sp. in Scott and Wilson, 2012. CCIS L2-245, Devonian 03, 02

UALVP19254 ventral surface of dorsal shield. Adrain and Wilson, 1994, p. 312, but reidentified as *Waengsjoeaspis* sp. in Scott and Wilson, 2012, p. 1246. CCIS L2-245, Devonian 03, 02

UALVP19255 head shield part with spine. Adrain and Wilson, 1994, p. 312. CCIS L2-245, Devonian 03, 02

UALVP19256 head shield. Adrain and Wilson, 1994, p. 312; reidentified as *Waengsjoeaspis* sp. in Scott and Wilson, 2014. CCIS L2-245, Devonian 03, 02

UALVP19257 part of dorsal head shield and partial body preserved mostly in ventral view; including Acanthodii. Adrain and Wilson, 1994; reidentified as *Waengsjoeaspis* sp. in Scott and Wilson, 2012, p. 1239, p. 1246. CCIS L2-245, Devonian 03, 02

UALVP19371 poorly preserved trunk and tail. Adrain and Wilson, 1994 p. 311. CCIS L2-245, Devonian 03, 02

UALVP23154 shield fragment. Adrain and Wilson, 1994, p. 312; reidentified as *Waengsjoeaspis* sp. in Scott and Wilson, 2014, p. 1246. *Polymerolepis whitei* according to Hanke *et al.* (2013); Wilson and Caldwell, 1998, Vol. 18(1):10–29, Descr. p. 20–21 Paratype for *Furcacauda fredhomae*. CCIS L2-245, Devonian 03, 02

UALVP23185 partial (plates). Adrain and Wilson, 1994, p. 312; reidentified as *Waengsjoeaspis* sp. in Scott and Wilson, 2014, p. 1246; Paratype for *Machairaspis serrata* Scott and Wilson, 2013, Fig. 3, B, p. 131; *Polymerolepis whitei* according to Hanke *et al.* (2013); Wilson and Caldwell, 1998 J. Vert. Paleo. Vol. 18(1):10–29, Descr. p. 20–21; Paratype for *Furcacauda fredhomae* Wilson and Caldwell, 1998. CCIS L2-245, Devonian 03, 02

UALVP23268 poorly preserved trunk and cornua. Adrain and Wilson, 1994, Descr. p. 311; Scott and Wilson, 2012, p. 1246. CCIS L2-245, Devonian 03, 02

UALVP32908 plate. Adrain and Wilson, 1994, Descr. p. 312; reidentified as *Waengsjoeaspis* sp. in Scott and Wilson, 2012, Descr. p. 1246. CCIS L2-245, Devonian 03, 03

UALVP32910 plates. Adrain and Wilson, 1994, Descr. p. 312; reidentified as *Waengsjoeaspis* sp. in Scott and Wilson, 2012, p. 1246. CCIS L2-245, Devonian 03, 03

UALVP32913 plates. Adrain and Wilson, 1994, Descr. p. 312; reidentified as *Waengsjoeaspis* sp. in Scott and Wilson, 2012, p, 1246. CCIS L2-245, Devonian 03, 03

UALVP32914 scales. Adrain and Wilson, 1994, Descr. p. 312; reidentified as *Waengsjoeaspis* sp. in Scott and Wilson, 2012, p. 1246. CCIS L2-245, Devonian 03, 03

UALVP32915 cephalic brim and right side of oral margin. Adrain and Wilson, 1994, Descr. p. 311; reidentified as *Waengsjoeaspis* sp. in Scott and Wilson, 2012, pp. 1239–1240. CCIS L2-245, Devonian 03, 04

UALVP32968 plate. Adrain and Wilson, 1994, Descr. p. 312; reidentified as *Waengsjoeaspis* sp. in Scott and Wilson, 2012, p. 1246. CCIS L2-245, Devonian 03, 04

UALVP32969 largely complete and well-preserved head showing inner surface of dorsal shield, Adrain and Wilson, 1994, fig. Fig. 8B, p. 311; Sahney and Wilson 2001 JVP 21(4):660–669 Fig. 2, C, p. 663, Fig 3, A, p. 664, as *Waengsjoeaspis nahanniensis* Fig. 4, C, p. 666; Fig. 5, C, p. 667. Visceral view of mostly complete head shield. Scott and Wilson, 2012, p. 1240, Fig. 3, D, p. 1241, as *Waengsjoeaspis platycornis* Scott and Wilson, 2012, paratype; Z-425, Unit 10, Drawer 5.

UALVP32976 plate, part & counterpart. Adrain and Wilson, 1994, Descr. p. 311; reidentified as *Waengsjoeaspis* sp. in Scott and Wilson, 2012, Descr. p. 1246. CCIS L2-245, Devonian 03, 04

UALVP32982 plate and scattered fragments. Adrain and Wilson, 1994, Descr. p. 312; reidentified as *Waengsjoeaspis* sp. in Scott and Wilson, 2012, JVP 32(6):1235–1253, Descr. p. 1246. CCIS L2-245, Devonian 03, 04

NOT PARATYPES OF _Waengsjoeaspis nahanniensis:_

UALVP 41643 Scott and Wilson, 2012, JVP 32(6):1235–1253, Descr. as a paratype for *Waengsjoeaspis nahanniensis* in Table 1,

p. 1239, but this was collected two years after Adrain and Wilson, 1994, was published. CCIS L2-245, Devonian 03, 06

UALVP 41747 Scott and Wilson, 2012, JVP 32(6):1235–1253, Descr. as a paratype for *Waengsjoeaspis nahanniensis* in Table 1, p. 1239, but this was collected two years after Adrain and Wilson, 1994, was published. CCIS L2-245, Devonian 03, 06

UALVP 42213 Scott and Wilson, 2012, JVP 32(6):1235–1253, Descr. as a paratype for *Waengsjoeaspis nahanniensis* in Table 1, p. 1239, but this was collected two years after Adrain and Wilson, 1994, was published. Biological Sciences Building Z-425 Unit 10, Drawer 4.

UALVP 52605 Scott and Wilson, 2012, JVP 32(6):1235–1253, Descr. as a paratype for *Waengsjoeaspis nahanniensis* in Table 1, p. 1239, but this was collected two years after Adrain and Wilson, 1994, was published. CCIS L2-245, Devonian 03, 09

Agnatha Haeckel, 1895
    Osteostraci Lankester, 1868
        Cornuata Janvier, 1985
            Family *incertae sedis*
UALVP19145 HOLOTYPE *Waengsjoeaspis platycornis* **Scott and Wilson, 2012**
Collector: Chatterton, Brian D. E., 1983.
Canada: Northwest Territories, Mackenzie Mountains. MOTH: Man on the Hill #1, Devonian-Devonian Early-Lochkovian, Road River Fm. Partial cephalic shield with intact left pectoral fin and articulated trunk scales. Scott and Wilson, 2012, Descr. p. 1240, Fig. 3, p. 1241, p. 1239, Table 1, p. 1239, partial cephalic shield also Paratype for *Waengsjoeaspis nahanniensis* (Fig. 7, B, p. 311); (on display in museum), Earth Sciences Museum, ESB B-01

    Scott, Bradley R. and Wilson, Mark V. H. 2012. A new species of *Waengsjoeaspis* (Cephalaspidomorpha, Osteostraci) from the early Devonian of Northwestern Canada, with a redescription of *W. nahanniensis* and implications for growth, variation, morphology, and phylogeny. Journal of Vertebrate Paleontology. Vol. 32(6):1235–1253.

PARATYPES:

UALVP19141 right half of margin of head shield orbits, and naso-hypophyseal region paratype for *Waengsjoeaspis nahanniensis*; Paratype for *Waengsjoeaspis platycornis* Scott and Wilson, 2012, Descr. p. 1240, CCIS L2-245, Devonian 03, 02

UALVP32969 visceral view of mostly complete head shield. Scott and Wilson, 2012, Descr. Table 1, p. 1239, Fig. 3, D, p. 1241; see also Adrain and Wilson, 1994, Fig. 8, B, also a paratype of *Waengsjoeaspis nahanniensis*. Biological Sciences Building Z-425, Unit 10, Drawer 5, L2-c

UALVP41477 complete dorsal head shield, preserved in visceral view. Scott and Wilson, 2012, Descr. Table 1, p. 1239, Fig. 3C, p. 1241. Biological Sciences Building Z-425, Unit 10, Drawer 5, L2-c

UALVP41663 (? Possible misquote of catalogue number?) Partial head shield with orbits, nasohypophyseal region, & pineal furrow preserved in dorsal view Scott and Wilson, 2012, Descr. Table 1, p. 1239, listed as Paratype p. 1240. However also listed as a paratype for both Acanthodii, *Tricuspicanthus gannitus* Blais, Hermus, and Wilson, 2015, p. 7. CCIS L2-245, Devonian 15, 19

UALVP41883 small head shield with trunk preserved in visceral view. Scott and Wilson, 2012, Descr. Table 1, p. 1239, Fig. 4, B, p. 1242 Scott and Wilson, 2012, p. 1240, p. 1248 CCIS L2-245, Devonian 03, 07

UALVP43086 (now UALVP55519, Double use of catalogue number) nearly complete dorsal head shield preserved in visceral view. Scott and Wilson, 2012, Descr. Table 1, p. 1239, Fig. 4, D, p. 1242. CCIS L2-245, Devonian 03, 09

UALVP44008 cornual processes and margin of head shield in ventral view. Scott and Wilson, 2012, Descr. Table 1, p. 1239. CCIS L2-245, Devonian 03, 08

UALVP52493 (This specimen had a catalogue number on it **UALVP41477** that was already assigned to another cata-logued specimen and was reassigned a new catalogue number

UALVP52493). Head shield in internal view, dermoskeleton not preserved. Scott and Wilson, 2012, Descr. Table 1, p. 1239, Fig. 4, C, p. 1242. CCIS L2-245, Devonian 03, 09

UALVP52495 partial head shield with fracture along right side and spreading of head shield across left side Scott and Wilson, 2012, fig. Fig. 4, A, p. 1242. CCIS L2-245, Devonian 03, 09

UALVP55519 (See UALVP43086 above, double use of catalogue number, formerly UALVP43086) Nearly complete dorsal head shield preserved in visceral view. Scott and Wilson, 2012, Descr. Table 1, p. 1239, Fig. 4, D, p. 1242. CCIS L2-245, Devonian 03, 09

Agnatha Haeckel, 1895
    Thelodonti Kiaer, 1932
        Furcacaudiformes Wilson and Caldwell, 1998
        Furcacaudidae Wilson and Caldwell, 1998
UALVP44912 HOLOTYPE *Canonia costulata* Märss, Wilson, and Thorsteinsson, 2002, p. 113
Collectors: IGCP 328 Party, 1994.
Canada: Nunavut, Cornwallis Island, RBBI*-94 34.5m, Devonian-Devonian Early-Lochkovian, Barlow Inlet Fm. Scale mounted on STUB130. Märss *et al.*, 2002, Descr. p. 113, & fig. Plate 1, fig. 15, p. 105. 1. Märss *et al.* 2006. Spec. Pap. Palaeo. 75: Descr. p. 118, & fig. Plate 24, Fig. 7, p. 119.

    Märss, Tiiu, Wilson, Mark V. H., and Thorsteinsson, Raymond 2002. New thelodont (Agnatha) and possible chondrichthyan (Gnathostomata) taxa established in the Silurian and Lower Devonian of the Canadian Arctic Archipelago. Proceedings of the Estonian Academy of Sciences Geology. Vol. 51(2): 87–122.

Agnatha, Haeckel, 1895
    Thelodontii Jaekel, 1911
        Furcacaudiformes Wilson and Caldwell, 1998, p. 12
        Furcacaudidae Wilson and Caldwell, 1998, p. 15
UALVP33027 Holotype *Cometicercus talimaaae* **Wilson and Caldwell, 1998, p. 15**

Canada: Northwest Territories, Mackenzie Mountains, MOTH: Man on the Hill #1, Devonian-Devonian Early-Lochkovian. Road River FM. plates, Wilson and Caldwell, 1998, Descr. p. 15–16, Fig. 4, incorrectly cited as UALVP33207, p. 15 CCIS L2-245, Devonian 15, 02

Wilson, Mark V. H., and Caldwell, Michael W. 1998. The Furcacaudiformes: a new order of jawless vertebrates with thelodont scales, based on articulated Silurian and Devonian fossils from northern Canada. Journal of Vertebrate Paleontology, 18(1):10–29.

Agnatha Haeckel, 1895
Thelodonti Jaekel, 1911
Furcacaudiformes Wilson and Caldwell, 1998
Furcacaudidae Wilson and Caldwell, 1998
UALVP32917 HOLOTYPE *Drepanolepis maerssae* Wilson and Caldwell, 1998, pp. 16–17.
Canada: Northwest Territories, Mackenzie Mountains MOTH: Man on the Hill #1 Devonian-Devonian Early-Lochkovian, Road River Fm. partial skeleton Wilson and Caldwell, 1998, Descr. p. 16–17, Fig. 5, a-d, p. 17 CCIS L2-245, Devonian 14, 05

Wilson, Mark V. H., and Caldwell, Michael W. 1998. The Furcacaudiformes: a new order of jawless vertebrates with thelodont scales, based on articulated Silurian and Devonian fossils from northern Canada. Journal of Vertebrate Paleontology Vol. 18(1):10–29.

Agnatha Haeckel, 1895
Thelodonti Jaekel, 1911
Furcacaudiformes Wilson and Caldwell, 1998, p. 12
Furcacaudidae Wilson and Caldwell, 1998, p. 15
UALVP33024 Holotype *Furcacauda fredholmae* Wilson and Caldwell, 1998, p. 22
Canada: Northwest Territories, Mackenzie Mountains, MOTH: Man on the Hill #1 Devonian-Devonian Early-Lochkovian, Road

River Fm. Laterally compressed individual, missing the head and most of the posterior extremeties of the caudal fin lobes. Wilson and Caldwell, 1998, Descr. pp. 21–22, & fig. Fig. 8, a–c, p. 22. CCIS L2-245, Devonian 15, 02

Wilson, Mark V. H., and Caldwell, Michael W. 1998. The Furcacaudiformes: a new order of jawless vertebrates with thelodont scales, based on articulated Silurian and Devonian fossils from northern Canada. Journal of Vertebrate Paleontology Vol. 18(1):10–29.

PARATYPES:

UALVP23154 shield fragment; Paratype for *Waengsjoeaspis nahanniensis* Adrain and Wilson, 1994, but reidentified as *Waengsjoeaspis* sp. in Scott and Wilson, 2014, p. 1246. Paratype for *Machairaspis serrata* Scott and Wilson, 2013, Fig. 3, B, p. 131. *Polymerolepis whitei* according to Hanke *et al.* 2013. Paratype for *Furcacauda fredholmae* Wilson and Caldwell, 1998, Descr. pp. 21–22 CCIS L2-245, Devonian 03, 02; CCIS L2-245, Devonian 08, 03

UALVP32417 fish Wilson and Caldwell, 1998, Descr. p. 21–22

UALVP32462 fish; on slab UALVP39086 Wilson and Caldwell, 1998, Descr. pp. 21–22 identified as *Furcacauda heintzae* on p. 20, identified as *Furcacauda fredholmae* on p. 22. CCIS L2-245, Devonian 14, 12

UALVP32949 scattered plate fragments Wilson and Caldwell, 1998, Descr. pp. 21–22. CCIS L2-245, Devonian 15, 02

UALVP38074 posterior half, temporary no. MB5 Wilson and Caldwell, 1998, Descr. pp. 21–22, Fig. 8, D, p. 22. CCIS L2-245, Devonian 15, 02

UALVP39085 posterior only; on back of slab containing UALVP32956; Wilson and Caldwell, 1998, Descr. pp. 21–22. CCIS L2-245, Devonian 14, 04

Agnatha Haeckel, 1895
  Thelodonti Jaekel, 1911
    Furcacaudiformes Wilson and Caldwell, 1998
      Furcacaudidae Wilson and Caldwell, 1998, p. 15

UALVP33023 HOLOTYPE *Sphenonectris turnerae* **Wilson and Caldwell, 1998, pp. 17–20.**
Canada: Northwest Territories, Mackenzie Mountains, MOTH: Man on the Hill #1. Devonian-Devonian Early-Lochkovian, Road River Fm. Plate Wilson and Caldwell, 1998, Descr. pp. 17–18, Fig. 6, A, C, D, p. 18. CCIS L2-245, Devonian 15, 02

Wilson, Mark V. H., and Caldwell, Michael W. 1998. The Furcacaudiformes: a new order of jawless vertebrates with thelodont scales, based on articulated Silurian and Devonian fossils from northern Canada. Journal of Vertebrate Paleontology Vol. 18(1):10–29.

PARATYPES:

UALVP23274 partial fish. Wilson and Caldwell, 1998, Descr. pp. 17–19. CCIS L2-245, Devonian 15, 02

UALVP23377 body parts. Wilson and Caldwell, 1998, Descr. pp. 17–19. CCIS L2-245, Devonian 15, 03

UALVP32920 collection of plates, scales, and scattered fragments. Portions of several over lapping heads, trunks, and tails. Wilson and Caldwell, 1998, Descr. pp. 17–19, & fig. Fig. 6, F, p. 18. CCIS L2-245, Devonian 03, 04

UALVP32950 scattered plate fragments. Wilson and Caldwell, 1998, Descr. pp. 17–19. CCIS L2-245, Devonian 15, 03

UALVP32954 plates. Wilson and Caldwell, 1998, Descr. pp. 17–19. CCIS L2-245, Devonian 15, 03

UALVP32960 plates. Wilson and Caldwell, 1998, Descr. pp. 17–19. CCIS L2-245, Devonian 15, 03

UALVP32961 scattered plates. Wilson and Caldwell, 1998, Descr. pp. 17–19 CCIS L2-245, Devonian 15, 03

UALVP32964 plate. Wilson and Caldwell, 1998, Descr. pp. 17–19. CCIS L2-245, Devonian 15, 03

UALVP32965 fish scales. Wilson and Caldwell, 1998, Descr. pp. 17–19. CCIS L2-245, Devonian 15, 03

UALVP32984 plates. Wilson and Caldwell, 1998, Descr. pp. 17–19. CCIS L2-245, Devonian 02, 03

UALVP33025 fish scales. Wilson and Caldwell, 1998, Descr. pp. 17–19. Measurements p. 19. CCIS L2-245, Devonian 15, 03
UALVP33026 plate. Wilson and Caldwell, 1998, Descr. pp. 17–19. CCIS L2-245, Devonian 15, 03
UALVP37149 tail. Wilson and Caldwell, 1998, Descr. pp. 17–19. CCIS L2-245, Devonian 15, 03
UALVP38075 middle 2/3s of the fish; no. MB3. Wilson and Caldwell, 1998, Descr. pp. 17–19. Measurements p. 19. CCIS L2-245, Devonian 15, 02

Agnatha Haeckel, 1895
Thelodonti Jaekel, 1911
Furcacaudiformes Wilson and Caldwell, 1998
Pezopallichthyidae Wilson and Caldwell, 1998
UALVP32994 HOLOTYPE *Pezopallichthys ritchiei* **Wilson and Caldwell, 1998**
Canada: Northwest Territories, Mackenzie Mountains, Silurian-Silurian Early-Wenlockian Late-Homerian, Road River Fm. skeleton. Wilson and Caldwell, 1998, Descr. pp. 12–15, & fig. Fig. 2, a, p. 13. CCIS L2-245, Silurian 09, 12
Wilson, Mark V. H., and Caldwell, Michael W. 1998. The Furcacaudiformes: a new order of jawless vertebrates with thelodont scales, based on articulated Silurian and Devonian fossils from northern Canada. Journal of Vertebrate Paleontology Vol. 18(1):10–29.
PARATYPES:
UALVP29922 Wilson and Caldwell, 1998, Descr. p. 12, L1; Biological Sciences Building Z-425, Unit 10, Drawer 2
UALVP29925 Wilson and Caldwell, 1998, Descr. p. 12 & fig. Fig. 2, g, p. 13, L1; Biological Sciences Building Z-425, Unit 10, Drawer 2
UALVP32991 plate. Wilson and Caldwell, 1998, Descr. p. 12, & fig. Fig. 2, c, p. 13. CCIS L2-245, Silurian 09, 12
UALVP32992 plate, part and counterpart. Wilson and Caldwell, 1998, Descr. p. 12, & fig. Fig. 2, a, p. 13. CCIS L2-245, Silurian 09, 12

UALVP32993 plate. Wilson and Caldwell, 1998, Descr. p. 12. CCIS L2-245, Silurian 09, 12

UALVP32995 plate. Wilson and Caldwell, 1998, Descr. p. 12, & fig. Fig. 2, g, p. 13.

UALVP32997 scales. Wilson and Caldwell, 1998, Descr. p. 12. CCIS L2-245, Silurian 09, 12

UALVP33001 plate and scattered fragments. Wilson and Caldwell, 1998, Descr. p. 12, & fig. Fig. 2, e, p. 13. CCIS L2-245, Silurian 09, 12

UALVP33002 plate. Wilson and Caldwell, 1998, Descr. p. 12. CCIS L2-245, Silurian 09, 12

UALVP33004 squamation. Wilson and Caldwell, 1998, Descr. p. 12, Fig. 2, F, p. 13

UALVP33005 plate and scales. Wilson and Caldwell, 1998, Descr. p. 12. CCIS L2-245, Silurian 09, 12

UALVP33007 plate. Wilson and Caldwell, 1998, Descr. p. 12. CCIS L2-245, Silurian 09, 12

UALVP33008 plate. Wilson and Caldwell, 1998, Descr. p. 12. CCIS L2-245, Silurian 09, 12

UALVP33009 scales. Wilson and Caldwell, 1998, Descr. p. 12, CCIS L2-245, Silurian 09, 12

Agnatha Haeckel, 1895
    Thelodonti Kiaer, 1932
        Loganelliiformes Turner, 1991
            Loganelliidae Karatajutè-Talimaa, 1997
UALVP43129 HOLOTYPE *Illoganellia colossea* **Märss, Wilson, and Thorsteinsson, 2002, p. 33**
IGCP 328 Party, 1994
Canada: Nunavut, Cornwallis Island, CP-94-TQ Cape Phillips Thorsteinsson Quarry, Silurian-Silurian Early-Wenlock Middle-Sheinwoodian, Cape Phillips. A patch of scales. Märss *et al.*, 2002, Plate II, fig. 5, p. 106, Descr. p. 94. Märss *et al.*, 2006, Spec. Pap. Paleo. 75, Descr. p. 33, & fig. Text-Fig. 16, C-D, p. 38; Text-Fig. 17, A-P, p. 39. CCIS L2-245, Silurian 13, 10

Märss, Tiiu, Wilson, Mark V. H., and Thorsteinsson, Raymond 2002. New thelodont (Agnatha) and possible chondrichthyan (Gnathostomata) taxa established in the Silurian and Lower Devonian of the Canadian Arctic Archipelago. Proceedings of the Estonian Academy of Sciences Geology. Vol. 51(2): 87–122.

Agnatha Haeckel, 1895
Thelodonti Kiaer, 1932
Loganelliiformes Turner, 1991
Loganelliidae Karatajutè-Talimaa, 1997
UALVP44658 HOLOTYPE *Loganellia exilis* **Märss, Wilson, and Thorsteinsson, 2002, p. 91**
IGCP 328 Party, 1994.
Canada: Nunavut, Baillie-Hamilton Island, BH-2–94 96.5 m, Silurian-Silurian Early-Wenlock Late-Homerian, Cape Phillips. Scale mounted on STUB120. Märss *et al.*, 2002, p. 91, Plate 1, Fig. 2. Märss *et al.*, 2006, Spec. Pap. Palaeo. 75. Fig. Plate 5, Fig. 17, p. 35.

Märss, Tiiu, Wilson, Mark V. H., and Thorsteinsson, Raymond 2002. New thelodont (Agnatha) and possible chondrichthyan (Gnathostomata) taxa established in the Silurian and Lower Devonian of the Canadian Arctic Archipelago. Proceedings of the Estonian Academy of Sciences Geology. Vol. 51(2): 87–122.

Agnatha Haeckel, 1895
Thelodonti Kiaer, 1932
Loganelliiformes Turner, 1991
Loganelliidae Karatajutè-Talimaa, 1997
UALVP39906 Cast of HOLOTYPE Cast of GSS #1137 at Royal Scottish Museum, Edinburgh. *Thelodus planus* **Traquair, 1898** now identified as *Loganellia plana* (Traquair, 1898)
United Kingdom: Scotland: East Ayrshire, Muirkirk, Seggholm Silurian-Silurian Late, Ludlovian, Slot Burn Formation. CCIS L2-245, Silurian 11, 01

Traquair, R. H. 1898. Report on fossil fishes collected by the Geological Survey of Scotland in the Silurian Rocks of the south of Scotland. Transactions of the Royal Society of Edinburgh Vol. 39:827–864.

Agnatha Haeckel, 1895
  Thelodonti Kiaer, 1932
    Loganelliiformes Turner, 1991
      Loganelliidae Karatajutè-Talimaa, 1997
Loganellia sulcata Märss, Wilson, and Thorsteinsson, 2002, p. 90
Collectors: Lindoe, L. Allan; Caldwell, Mike W., 1994.
Canada: Nunavut, Cornwallis Island, CP-94-TQ Cape Phillips Thorsteinsson Quarry, Silurian-Silurian Early-Wenlock Middle-Sheinwoodian, Cape Phillips. Well-preserved anterior articulated specimen.

Märss, Tiiu, Wilson, Mark V. H., and Thorsteinsson, Raymond 2002. New thelodont (Agnatha) and possible chondrichthyan (Gnathostomata) taxa established in the Silurian and Lower Devonian of the Canadian Arctic Archipelago. Proceedings of the Estonian Academy of Sciences Geology. Vol. 51(2): 87–122.

PARATYPES:

UALVP43150 Märss *et al.* 2002, Descr. pp. 90–91, & fig. Plate 1, Fig. 1, p. 105 Glass slide 907 T1 (**UALVP43140** Plate 2, Fig. 3 is the counterpart to **UALVP43150**). Märss *et al.* 2006. Spec. Pap. Pal. 75, Descr. p. 21, & fig. Plate 2, fig. 2; 4–6, Text-figs. 9, 10 a–Z, Text-Fig. 11, B, p. 28. CCIS L2-245, Silurian 13, 11

UALVP44565 scale mounted on STUB118. Märss *et al.* 2002, 51(2):88–120, Plate 1, fig. 1, p. 105. Märss *et al.* 2006. Spec. Pap. Pal. 75 fig. Plate 3, Fig. 10, p. 27.

Agnatha Haeckel, 1895
  Thelodonti Kiaer, 1932
    Loganelliiformes Turner, 1991

Nunavutiidae Märss, Wilson, and Thorsteinsson, 2002, p. 95.

UALVP44741 HOLOTYPE *Nunavutia fasciata* **Märss, Wilson, and Thorsteinsson, 2002, p. 95**
Collectors: IGCP 328 Party, 1994.
Canada: Nunavut, Baillie-Hamilton Island, BH-2–94 220m Cape Phillips Thorsteinsson Quarry, Silurian-Silurian Late-Ludlow-Gorstian, Cape Phillips Fm. Scale mounted on STUB122. Märss *et al.*, 2002, 51(2):87–120, Descr. p. 95, & fig. Plate I, Fig. 3. Märss *et al.*, 2006. Spec. Pap. Paleo. 75, Descr. p. 34

Märss, Tiiu, Wilson, Mark V. H., and Thorsteinsson, Raymond 2002. New thelodont (Agnatha) and possible chondrichthyan (Gnathostomata) taxa established in the Silurian and Lower Devonian of the Canadian Arctic Archipelago. Proceedings of the Estonian Academy of Sciences Geology. Vol. 51(2): 87–122.

Agnatha Haeckel, 1895
> **Thelodontii Jaekel, 1911**
>> **Phlebolepidiformes Berg, 1937**
>>> **Katoporodidae Märss, Wilson, and Thorsteinsson, 2002**

UALVP45052 HOLOTYPE *Overia adraini* **Soehn, Märss, Caldwell, and Wilson, 2001, p. 656**
Collector: Chatterton, Brian D. E., 1979.
Canada: Northwest Territories, Avalanche Lake Section AV2 256A, Silurian-Silurian Early-Wenlock, Sheinwoodian-Homerian, Delorme Fm. Scale. Soehn *et al.* 2001, Descr. p. 656, & fig Fig. 4, G, p. 657; Märss *et al.*, 2006. Spec. Pap. Palaeo. 75, Descr. p. 67, from 256 meters.

Soehn, Kenneth L., Märss, Tiiu, Caldwell, Michael W., and Wilson, Mark V. H. 2001. New and biostratigraphically useful thelodonts from the Silurian of the Mackenzie mountains, Northwest Territories, Canada. Journal of Vertebrate Paleontology Vol. 21(4): 651–659

PARATYPES:

UALVP43014 scale (as *Loganellia*) Figured Plate 1, Fig. 5. Soehn *et al.* 2000; Descr. p. 656. Soehn *et al.* 2001, Reserved for scale from Avalanche Lake sections.

UALVP43015 scale Figured Plate 1, Fig. 6. Soehn *et al.* 2000 (as *Loganellia* sp. nov.); Soehn *et al.* 2001, fig. Fig. 4, Descr. p. 657, Reserved for scale from Avalanche Lake sections.

UALVP43016 scale Figured Plate 1, Fig. 7. Soehn *et al.* 2000 (as *Loganellia* sp. nov.); Soehn *et al.* 2001, Descr. p. 656, Reserved for scale from Avalanche Lake sections

UALVP43017 scale Figured (as *Loganellia*) Plate 1, Fig. 8. Soehn *et al.* 2000 (as *Loganellia* sp. nov.); Soehn *et al.* 2001, Descr. p. 656, Reserved for scale from Avalanche Lake sections

UALVP45051 scale. Soehn *et al.* 2001, Descr. p. 656 & fig. Fig. 4, A, p. 657

UALVP45053 scale. Soehn *et al.* 2001, Descr. p. 656 & fig. Fig. 4, I, p. 657

UALVP45054 scale. Soehn *et al.* 2001, Descr. p. 656 & fig. Fig. 4, B, p. 657, 242 m

UALVP45055 scale. Soehn *et al.* 2001, Descr. p. 656 & fig. Fig. 4, C, p. 657

UALVP45056 scale. Soehn *et al.* 2001, Descr. p. 656 & fig. Fig. 4, E, p. 657

UALVP45057 scale. Soehn *et al.* 2001, Descr. p. 656 & fig. Fig. 4, H, p. 657

UALVP45058 scale. Soehn *et al.* 2001, Descr. p. 656 & fig. Fig. 4, F, p. 657

UALVP45059 scale. Soehn *et al.* 2001, Descr. p. 656 & fig. Fig. 4, J, p. 657

UALVP45070 scale. Soehn *et al.* 2001, fig. Fig. 4, K, p. 657. [Note: although a paratype, this specimen number was left off from the PARATYPES listed on p. 656].

UALVP45071 scale. Soehn *et al.* 2001, fig. Fig. 4, L, p. 657. [Note: although a paratype, this specimen number was left off from the PARATYPES listed on p. 656].

Agnatha Haeckel, 1895
  Thelodonti Kiaer, 1932
    Phlebolepidiformes Berg, 1937
      Katoporodidae Märss, Wilson, and Thorsteinsson, 2002, p. 98
UALVP44709 HOLOTYPE *Katoporodus gemellus* Märss, Wilson, and Thorstcinsson, 2002, p. 99. Now identified as *Trimerolepis gemella* (Märss, Wilson, and Thorsteinsson, 2002, p. 105)
Collectors: IGCP 328 Party, 1994.
Canada: Nunavut, Cornwallis Island, RBBI*-94 34.5m, Devonian-Devonian Early-Lochkovian. Barlow Inlet Fm. Scale mounted on STUB121. Märss *et al.*, 2002, 51(2):87–120, Descr. p. 99, & fig. Plate 1, Figs. 7, 8. p. 105. Märss *et al.*, 2006, Spec. Pap. Palaeo. 75, Descr. p. 64, & fig. Plate 13, Fig. 6, p. 65; Placed in *Trimerolepis* Märss *et al.* 2007, "Agnathi" II Thelodonti Vol. 1B, Descr. p. 82, & fig. Fig. 81, M, N, p. 79
    Märss, Tiiu, Wilson, Mark V. H., and Thorsteinsson, Raymond 2002. New thelodont (Agnatha) and possible chondrichthyan (Gnathostomata) taxa established in the Silurian and Lower Devonian of the Canadian Arctic Archipelago. Proceedings of the Estonian Academy of Sciences Geology. Vol. 51(2): 87–122.

Agnatha Haeckel, 1895
  Thelodonti Kiaer, 1932
    Phlebolepidiformes Berg, 1937
      Katoporodidae Märss, Wilson, and Thorsteinsson, 2002, p. 98
UALVP44923 HOLOTYPE *Katoporodus serratus* Märss, Wilson, and Thorsteinsson, 2002, p. 105. Now identified as *Trimerolepis serrata* (Märss, Wilson, and Thorsteinsson, 2002, p. 105)
Collectors: IGCP 328 Party, 1994.
Canada: Nunavut, Cornwallis Island, RBBI*-94 34.5m, Devonian-Devonian Early-Lochkovian. Barlow Inlet Fm. Scale mounted on

STUB130. Märss *et al.*, 2002, Descr. p. 99, & fig. Plate 1, Figs. 5, 6, p. 105. Märss *et al.*, 2006, Spec. Pap. Palaeo. 75, Descr. p. 60, & fig. Plate 12, Fig. 14, p. 61; Placed in *Trimerolepis* Märss *et al.* 2007, "Agnathi" II Thelodonti Vol. 1B, Descr. p. 82, & fig. Fig. 81, O, P, p. 79.

Märss, Tiiu, Wilson, Mark V. H., and Thorsteinsson, Raymond 2002. New thelodont (Agnatha) and possible chondrichthyan (Gnathostomata) taxa established in the Silurian and Lower Devonian of the Canadian Arctic Archipelago. Proceedings of the Estonian Academy of Sciences Geology. Vol. 51(2): 87–122.

PARATYPE:

UALVP44707 Scale mounted on STUB121. Märss *et al.*, 2006, Spec. Pap. Palaeo. 75. fig. Plate 12, Fig. 12, p. 61. Märss *et al.* 2002, fig. Plate 1, Fig. 6, p. 105

Agnatha Haeckel, 1895
Thelodontii Jaekel, 1911
Phlebolepidiformes Berg, 1937
Phlebolepididae Berg, 1940
UALVP55589 HOLOTYPE *Erepsilepis margaritifera* **Märss, Wilson, and Thorsteinsson, 2002**
IGCP 328 Party, 1994
Canada: Nunavut, Cornwallis Island, Cape Phillips Thorsteinsson Quarry Cornwallis Island, Silurian-Silurian Early-Wenlock Middle-Sheinwoodian, Cape Phillips. Märss *et al.*, 2002, Descr. p. 98, & fig. Plate 3, Fig. 1, p. 107; Fig. 1, f, p. 92. Märss *et al.*, 2006, Spec. Pap. Palaeo. 75. fig. Plate 11, 1–11, p. 58, Text-Fig. 25, p. 60. CCIS L2-245, Silurian 13, 10

Märss, Tiiu, Wilson, Mark V. H., and Thorsteinsson, Raymond 2002. New thelodont (Agnatha) and possible chondrichthyan (Gnathostomata) taxa established in the Silurian and Lower Devonian of the Canadian Arctic Archipelago. Proceedings of the Estonian Academy of Sciences Geology Vol. 51(2): 87–122.

PARATYPE:

UALVP43115 Märss *et al.*, 2002. Descr. p. 98, & fig. Plate 3, Fig. 1, p. 107; Fig. 1, f, p. 92. Märss *et al.*, 2006, Spec. Pap. Palaeo. 75,

Descr. p. 53, & fig. Plate 11, 1–11, p. 58, Text-Fig. 25, p. 60. CCIS L2-245, Silurian 13, 10

Agnatha Haeckel, 1895
  Thelodonti Kiaer, 1932
    Shieliiformes Märss, Wilson, and Thorsteinsson, 2002, p. 95
      Shieliidae Märss, Wilson, and Thorsteinsson, 2002, p. 95
UALVP44939 HOLOTYPE *Paralogania readbayensis* Märss, Wilson, and Thorsteinsson, 2002, p. 97
Collectors: IGCP 328 Party, 1994.
Canada: Nunavut, Cornwallis Island, RBBI*-94 61.5m, Devonian-Devonian Early-Lochkovian, Barlow Inlet Fm. Scale mounted on STUB131. Märss *et al.*, 2002, 51(2):87–120, Descr. p. 97, fig. Plate I, fig. 4, p. 105. Märss *et al.*, 2006, Spec. Pap. Palaeo. 75. fig. Plate 10, Fig. 18, p. 55
Märss, Tiiu, Wilson, Mark V. H., and Thorsteinsson, Raymond 2002. New thelodont (Agnatha) and possible chondrichthyan (Gnathostomata) taxa established in the Silurian and Lower Devonian of the Canadian Arctic Archipelago. Proceedings of the Estonian Academy of Sciences Geology Vol. 51(2): 87–122.

Agnatha Haeckel, 1895
  Thelodonti Kiaer, 1932
    Shieliiformes Märss, Wilson, and Thorsteinsson, 2002, p. 95
      Shieliidae Märss, Wilson, and Thorsteinsson, 2002, p. 95
UALVP39908 Cast of HOLOTYPE *Thelodus taiti* Stetson, 1931, p. 143, cast of GSS 3903, T2751 at Royal Scottish Museum, Edinburgh, now *Sheilia taiti* (Stetson, 1931)
United Kingdom: Scotland: East Ayrshire, Muirkirk, Seggholm Late Silurian, Slot Burn Formation. Stetson, 1931, Descr. p. 143–145, & fig. Fig. 2, B, C, p. 144. Fig. 3, p. 146. CCIS L2-245, Silurian 11, 01

Stetson, H. C. 1931. Studies on the morphology of the Heterostraci Journal of Geology Vol. 39(2):141–154.

Agnatha Haeckel, 1895
    Thelodontii Jaekel, 1911
        Thelodontiformes Kiaer, 1932
            Archipelepididae Märss *IN*: Soehn, Märss, Caldwell, and Wilson, 2001

UALVP32990 HOLOTYPE. *Archipelepis turbinata* Soehn, Märss, Caldwell, and Wilson, 2001, p. 652
Canada: Northwest Territories, Avalanche Lake, AV1a, Silurian-Silurian Early-Wenlock, Delorme Fm. Whole fish with scales Soehn *et al.* 2001, Descr. pp. 652–653, & fig. Fig. 2, A, p. 654; from 515 m (talus). CCIS L2-245, Silurian 9, 12

Soehn, Kenneth L., Märss, Tiiu, Caldwell, Michael W., and Wilson, Mark V. H. 2001. New and biostratigraphically useful thelodonts from the Silurian of the Mackenzie mountains, Northwest Territories, Canada Journal of Vertebrate Paleontology Vol. 21(4): 651–659

PARATYPES:
UALVP15797 articulated. Soehn *et al.* 2001, Descr. pp. 652–653, & fig. Fig. 2, B, p. 654. CCIS L2-245, Silurian 9, 12

UALVP32990 whole fish with scales. Soehn *et al.* 2001, Descr. pp. 652–653, & fig. Fig. 2, A, p. 654; from 515 m (talus). CCIS L2-245, Silurian 9, 12

UALVP43022 scale Figured Plate 1, Fig. 13. Soehn *et al.* 2000 (as?*Thelodus hoskinsi* Giffin, 1979); Descr. p. 652. Soehn *et al.* 2001, Reserved for scale from Avalanche Lake sections, figured in Soehn *et al.*

UALVP43023 scale Figured Plate 1, Fig. 14. Soehn *et al.* 2000 (as?*Thelodus hoskinsi* Giffin, 1979); Soehn *et al.* 2001, Descr. p. 652

UALVP43024 scale Figured Plate 1, Fig. 15. Soehn *et al.* 2000; Soehn *et al.* 2001, fig. Fig. 3, D, p. 655.

UALVP43025 scale Figured Plate 1, Fig. 15. Soehn *et al.* 2000; Soehn *et al.*, 2001, fig. Fig. 3, C, p. 655.

UALVP43060 scale. Soehn *et al.* 2001, Descr. p. 652
UALVP43061 scale. Soehn *et al.* 2001, Descr. p. 652
UALVP45021 fragment of caudal peduncle and base of tail. Soehn *et al.* 2001, fig. Fig. 2, F, G, p. 654 talus about 460 m. CCIS L2-245, Silurian 09, 12
UALVP45062 scale. Soehn *et al.* 2001, Descr. p. 652, & fig. Fig. 3, E, p. 655
UALVP45063 scale. Soehn *et al.* 2001, Descr. p. 652, & fig. Fig. 3, F, p. 655
UALVP45064 scale. Soehn *et al.* 2001, Descr. p. 652, & fig. Fig. 3, G, p. 655
UALVP45065 scale. Soehn *et al.* 2001, Descr. p. 652, & fig. Fig. 3, H, p. 655
UALVP45066 scale. Soehn *et al.* 2001, Descr. p. 652, & fig. Fig. 3, I, p. 655
UALVP45067 scale. Soehn *et al.* 2001, Descr. p. 652, & fig. Fig. 3, J, p. 655, 10 m
UALVP45068 scale. Soehn *et al.* 2001, Descr. p. 652, & fig. Fig. 3, K, p. 655 10 m
UALVP45069 scale. Soehn *et al.* 2001, Descr. p. 652, & fig. Fig. 3, L, p. 655 10 m

Agnatha Haeckel, 1895
   Thelodonti Kiaer, 1932
      Thelodontiformes Kiaer, 1932
         Barlowodidae Märss, Wilson, and Thorsteinsson, 2002, p. 113
UALVP44687 HOLOTYPE. *Barlowodus excelsus* **Märss, Wilson, and Thorsteinsson, 2002, p. 114**
Collectors: IGCP 328 Party, 1994.
Canada: Nunavut, Cornwallis Island, RBBI*-94 34.5m, Devonian-Devonian Early-Lochkovian. Barlow Inlet Fm. Scale mounted on STUB121. Märss *et al.*, 2002, Descr. p. 114, & fig. Plate 1, Fig. 16, p. 105. Märss *et al.*, 2006. Spec. Pap. Palaeo. 75: fig. Plate 25, Fig. 11, p. 123.

Märss, Tiiu, Wilson, Mark V. H., and Thorsteinsson, Raymond 2002. New thelodont (Agnatha) and possible chondrichthyan (Gnathostomata) taxa established in the Silurian and Lower Devonian of the Canadian Arctic Archipelago. Proceedings of the Estonian Academy of Sciences Geology. Vol. 51(2): 87–122.

Agnatha Haeckel, 1895
Thelodonti Kiaer, 1932
Thelodontiformes Kiaer, 1932
Barlowodidae Märss, Wilson, and Thorsteinsson, 2002, p. 113
UALVP44955 HOLOTYPE. *Barlowodus floralis* Märss, Wilson, and Thorsteinsson, 2002, p. 125
Collectors: IGCP 328 Party, 1994.
Canada: Nunavut, Cornwallis Island, RBBI*-94 61.5m, Devonian-Devonian Early-Lochkovian, Barlow Inlet Fm. Scale mounted on STUB131. Märss *et al.* 2002, Descr. p. 114, & fig. Plate 1, Fig. 17, p. 105. Märss *et al.*, 2006. Spec. Pap. Palaeo. 75, Descr. p. 125, & fig. Plate 26, Fig. 5, p. 127.

Märss, Tiiu, Wilson, Mark V. H., and Thorsteinsson, Raymond 2002. New thelodont (Agnatha) and possible chondrichthyan (Gnathostomata) taxa established in the Silurian and Lower Devonian of the Canadian Arctic Archipelago. Proceedings of the Estonian Academy of Sciences Geology. Vol. 51(2): 87–122.

Agnatha Haeckel, 1895
Thelodonti Kiaer, 1932
Thelodontiformes Kiaer, 1932
Barlowodidae Märss, Wilson, and Thorsteinsson, 2002, p. 113
UALVP44700 HOLOTYPE. *Barlowodus tridens* Märss, Wilson, and Thorsteinsson, 2002, p. 115)
Collectors: IGCP 328 Party, 1994.
Canada: Nunavut, Cornwallis Island, RBBI*-94 34.5m, Devonian-Devonian Early-Lochkovian. Barlow Inlet Fm. Scale mounted on

STUB121. Märss *et al.*, 2002, Descr. p. 115, & fig. Plate 1, Fig. 18, p. 105. Märss *et al.*, 2006. Spec. Pap. Palaeo. 75: Descr. p. 126 & Fig. Plate 25, Fig. 6, p. 123.

Märss, Tiiu, Wilson, Mark V. H., and Thorsteinsson, Raymond 2002. New thelodont (Agnatha) and possible chondrichthyan (Gnathostomata) taxa established in the Silurian and Lower Devonian of the Canadian Arctic Archipelago. Proceedings of the Estonian Academy of Sciences Geology. Vol. 51(2): 87–122.

Agnatha Haeckel, 1895
    Thelodonti Kiaer, 1932
        Thelodontiformes Kiaer, 1932
            Boothialepididae Märss, 1999
UALVP44723 HOLOTYPE *Boothialepis thorsteinssoni* **Märss, 1999, p. 1083**
Collectors: IGCP 328 Party, 1994.
Canada: Nunavut, Cornwallis Island, RBBI*-94 34.5m, Devonian-Devonian Early-Lochkovian. Barlow Inlet Fm. Scale mounted on STUB122. Märss, 1999, Descr. 1083. Märss *et al.*, 2002. Proc. Est. Acad. Sci. Geol. 51(2):87–120, Descr. p. 95, & fig. Plate I, Fig. 3. Märss *et al.,* 2006. Spec. Pap. Paleo. 75, Descr. p. 34

Märss, Tiiu 1999. A new thelodont from the Upper Silurian or Lower Devonian of the Boothia Peninsula, Arctic Canada. Palaeontology 42:1079–1099.
<u>**PARATYPE:**</u>
UALVP45990 Scale in thin section. Märss, 1999, Descr. 1083. Märss *et al.,*2006. Spec. Pap. Palaeo. 75 fig. Text-Fig. 49, B, p. 100. CCIS L2-245, Silurian 11, 07

Agnatha Haeckel, 1895
    Thelodontii Jaekel, 1911
        Thelodontiformes Kiaer, 1932
            Coelolepididae Pander, 1856
UALVP55799 (formerly UALVP43141) HOLOTYPE *Thelodus inauditus* **Märss, Wilson, and Thorsteinsson, 2002, p. 101**

Catalogue number UALVP43141 used twice as mammal and agnathan.

Collectors: Lindoe, L. Allan, and Caldwell, Mike W., 1994.

Canada: Nunavut, Cornwallis Island, CP-94-TQ Cape Phillips Thorsteinsson Quarry, Silurian-Silurian Early-Wenlock-Sheinwood, Cape Phillips Fm. Märss *et al.*, 2002, Descr. p. 101, & fig. Plate 1, Fig. 9, p. 105; Plate 3, Fig. 3, p. 107. Märss *et al.*, 2006. Spec. Pap. Palaeo. 75. fig. Text-Fig. 34, A-O, p. 74; Text-Fig. 35, A, B, p. 75. CCIS L2-245, Silurian 13, 07

Märss, Tiiu, Wilson, Mark V. H., and Thorsteinsson, Raymond. 2002. New thelodont (Agnatha) and possible chondrichthyan (Gnathostomata) taxa established in the Silurian and Lower Devonian of the Canadian Arctic Archipelago. Proceedings of the Estonian Academy of Sciences Geology 51(2): 87–122.

Agnatha Haeckel, 1895
  Thelodonti Kiaer, 1932
    Thelodontiformes Kiaer, 1932
      Lanarkiidae Obruchev, 1949
UALVP43123 HOLOTYPE *Phillipsilepis cornuta* **Märss, Wilson, and Thorsteinsson, 2002, p. 102**

Collector: Lindoe, L. Allan, 1994.

Canada: Nunavut, Cornwallis Island, CP-94-TQ Cape Phillips Thorsteinsson Quarry, Silurian-Silurian Early-Wenlock Middle-Sheinwoodian, Cape Phillips Fm. Large fragment without anterior part of head & without fins. Counterpart to UALVP43114. Märss *et al.*, 2002, Descr. p. 102, & fig. Plate 4, Fig. 5, p. 108; Fig. 1(h), p. 93. Märss *et al.*,2006. Spec. Pap. Palaeo. 75, Descr. p. 80, & fig. Text.-Fig. 43, A–P, p. 89; Plate 17, Figs. 1–13, p. 91, Text.-Fig. 44, A–B, p. 92. CCIS L2-245, Silurian 13, 10

Märss, Tiiu, Wilson, Mark V. H., and Thorsteinsson, Raymond 2002. New thelodont (Agnatha) and possible chondrichthyan (Gnathostomata) taxa established in the Silurian and Lower Devonian of the Canadian Arctic Archipelago. Proceedings of the Estonian Academy of Sciences Geology. Vol. 51(2): 87–122.

Agnatha Haeckel, 1895
  Thelodonti Kiaer, 1932
    Thelodontiformes Kiaer, 1932
      Nikoliviidae Karatajutè-Talimaa, 1978
UALVP44743 HOLOTYPE *Chattertonodus cometoides* Märss,
Wilson, and Thorsteinsson, 2002, p. 110
Collectors: IGCP 328 Party, 1994.
Canada: Nunavut, Cornwallis Island, RBBI-94 59m, Silurian-
Silurian Late-Pridoli, Barlow Inlet Fm.
Scale mounted on STUB122. Märss, *et al.*, 2002, Descr. p. 110, &
fig. Plate I, Fig. 13, p. 105. Märss *et al.*, 2006. Spec. Pap. Palaeo. 75,
Descr. p. 105, & fig. Plate 21, Fig. 12, p. 107
  **Märss, Tiiu, Wilson, Mark V. H., and Thorsteinsson,
Raymond 2002.** New thelodont (Agnatha) and possible chon-
drichthyan (Gnathostomata) taxa established in the Silurian and
Lower Devonian of the Canadian Arctic Archipelago. Proceedings
of the Estonian Academy of Sciences Geology. Vol. 51(2): 87–122.

Agnatha Haeckel, 1895
  Thelodonti Kiaer, 1932
    Thelodontiformes Kiaer, 1932
      Nikoliviidae Karatajutè-Talimaa, 1978
UALVP44706 HOLOTYPE *Nikolivia auriculata* Märss, Wilson,
and Thorsteinsson, 2002, p. 109
Collectors: IGCP 328 Party, 1994.
Canada: Nunavut, Cornwallis Island, RBBI*-94 34.5m, Devonian-
Devonian Early-Lochkovian. Barlow Inlet Fm. Scale mounted on
STUB121. Märss *et al.*, 2002, Descr. p. 109, fig. Plate 1, Figs. 11, 12,
p. 105. Märss *et al.*, 2006. Spec. Pap. Palaeo. 75: fig. Plate 21, Fig.
2, p. 107.
  **Märss, Tiiu, Wilson, Mark V. H., and Thorsteinsson,
Raymond 2002.** New thelodont (Agnatha) and possible chon-
drichthyan (Gnathostomata) taxa established in the Silurian and
Lower Devonian of the Canadian Arctic Archipelago. Proceedings
of the Estonian Academy of Sciences Geology. Vol. 51(2): 87–122.

Agnatha Haeckel, 1895
   Thelodonti Kiaer, 1932
      Thelodontiformes Kiaer, 1932
         Talivaliidae Märss, Wilson, and Thorsteinsson, 2002, p. 113

UALVP44969 HOLOTYPE *Glacialepis corpulenta* Märss, Wilson, and Thorsteinsson, 2002, p. 111
Collectors: IGCP 328 Party, 1994.
Canada: Nunavut, Cornwallis Island, Read Bay RBSB-94 66.0 m, Devonian-Devonian Early-Lochkovian, Sophia Lake Fm. Scale mounted on STUB132. Märss *et al.*, 2002. Descr. p. 111, & fig. Plate 1, fig. 14, p. 105. Märss *et al.*, 2006. Spec. Pap. Palaeo. 75: Descr. p. 110, & fig. Plate 22, Fig. 16, p. 109.

Märss, Tiiu, Wilson, Mark V. H., and Thorsteinsson, Raymond 2002. New thelodont (Agnatha) and possible chondrichthyan (Gnathostomata) taxa established in the Silurian and Lower Devonian of the Canadian Arctic Archipelago. Proceedings of the Estonian Academy of Sciences Geology. Vol. 51(2): 87–122.

Placodermi M'Coy, 1848
   Arthrodira Woodward, 1891
      Pachyosteina Stensiö, 1944
         Dinichthyoidea Denison, 1978
            Dinichthyidae Newberry, 1885

UALVP49099 CAST OF HOLOTYPE, cast of MMMN V2285 *Eastmanosteus lundarensis* Hanke, Stewart, and Lammers, 1996
Canada: Manitoba, 15 km northeast of Lundar, Lily Bay (East) Quarry, 50°46'N, 98°13'W. Lily Bay Quarry. Middle Devonian: Elm Point and Winnipegosis Formations. Almost complete articulated cranial roof. CCIS L2–245, Devonian 05, 02

Hanke, G. F., Stewart, K. W., and Lammers, G. E. 1996. *Eastmanosteus lundarensis* sp. nov. from the Middle Devonian Elm Point and Winnipegosis Formations of Manitoba. Journal of Vertebrate Paleontology Vol. 16(4):606–616.

Placodermi M'Coy, 1848
  Arthrodira Woodward, 1891
    Pachyosteina Stensiö, 1944
      Dinichthyoidea Denison, 1978
        Dinichthyidae Newberry, 1885
UALVP49100 CAST OF HOLOTYPE, cast of MMMN V2241
Squamatognathus steeprockensis Hanke, Stewart, and Lammers, 1996
Canada: Manitoba, near Steep Rock, LaFarge Quarry, 51°25′N, 98°40′W. LaFarge Quarry. Devonian-Devonian Middle-Eifelian: Elm Point Formation. Inferognathal. CCIS L2–245, Devonian 05, 02

> Hanke, G. F., Stewart, K. W., and Lammers, G. E. 1996. *Squamatognathus steeprockensis* gen. et sp. nov., an arthrodire inferognathal from the Middle Devonian Elm Point Formation of Manitoba. Journal of Vertebrate Paleontology. Vol. 16(4):617–622.

Chondrichthyes Huxley, 1880
  Elasmobranchii Bonaparte, 1838
UALVP43408 HOLOPTYPE *Aethelamia elusa* Hanke, 2001, p. 159
Collector: Lindoe, L. Allan, 1998. Canada: Northwest Territories, Mackenzie Mountains, MOTH: Man on the Hill #1, Devonian-Devonian Early-Lochkovian, Delorme Group. Dorsoventrally flattened mostly complete body; missing parts of rostrum and tail. Hanke, 2001, Descr. p. 159–183, p. 434, Fig. 43, p. 163, Fig. 44, p. 165, Fig. 45, 1–6, p. 167. CCIS L2-245, Devonian 14, 08

> Hanke, Gavin F. 2001. Comparison of an Early Devonian Acanthodian and Putative Chondrichthyan assemblage using both isolated and articulated remains from the Mackenzie Mountains, with a cladistic analysis of Early Gnathostomes. Ph.D. thesis Department of Biological Sciences, University of Alberta, Edmonton, Alberta, Canada 566 pp.

PARATYPES:

UALVP32418 small, near complete fish with jaws. Hanke, 2001, Descr. p. 159, Fig. 48.1–6, p. 177. CCIS L2-245, Devonian 05, 03

UALVP32844 scattered plate fragments. Hanke, 2001, Descr. p. 159. CCIS L2-245, Devonian 14, 08

UALVP32967 scattered scales. Hanke, 2001, Descr. p. 159. CCIS L2-245, Devonian 14, 08

UALVP32981 patch of scales. Hanke, 2001, Descr. p. 159. CCIS L2-245, Devonian 14, 08

UALVP39072 miscellaneous scales Type 1S scales. Hanke and Wilson, 1997. Ichthyolith Issues Special Publication 2. IGCP 406 p. 15. Hanke, 2001, Descr. p. 159. CCIS L2-245, Devonian 14, 08

UALVP41506 few scattered scales. Hanke, 2001, Descr. p. 159. CCIS L2-245, Devonian 14, 08

UALVP41553 scattered scales of varying size. Hanke, 2001, Descr. p. 159. CCIS L2-245, Devonian 14, 08

UALVP41696 partial body, medial view; possible lateral line and leading edge scales. Hanke, 2001, Descr. p. 159. CCIS L2-245, Devonian 14, 08

UALVP41792 scattered scales and spine bits. Hanke, 2001, Descr. p. 159. CCIS L2-245, Devonian 14, 08

UALVP41993 uncertain, in prep. Hanke, 2001, Descr. p. 159. CCIS L2-245, Devonian 14, 10

UALVP41996 scattered scales, a spine, and a patch of scales. Hanke, 2001, Descr. p. 159. CCIS L2-245, Devonian 14, 08

UALVP42153 two parts. Hanke, 2001, Descr. p. 159. CCIS L2-245, Devonian 04, 05

UALVP42164 few isolated scales. Hanke, 2001, Descr. p. 159. CCIS L2-245, Devonian 14, 08

UALVP42256 Hanke, 2001, Descr. p. 159. CCIS L2-245, Devonian 03, 12

UALVP42277 scattered scales, spines, and tooth whorls. Hanke, 2001, Descr. p. 159. CCIS L2-245, Devonian 14, 02

UALVP42277.1 (from block UALVP42277) isolated scale (still on SEM stub). Hanke, 2001, Descr. p. 159

UALVP42277.2 (from block UALVP42277) isolated scale. Hanke, 2001, Descr. p. 159

UALVP42277.3 (from block UALVP42277) isolated scale. Hanke, 2001, Descr. p. 159

UALVP42277.4 (from block UALVP42277) isolated scale. Hanke, 2001, Descr. p. 159

UALVP42277.5 (from block UALVP42277) isolated scale. Hanke, 2001, Descr. p. 159

UALVP42277.6 (from block UALVP42277) isolated scale. Hanke, 2001, Descr. p. 159

UALVP42277.7 (from block UALVP42277) isolated scale. Hanke, 2001, Descr. p. 159

UALVP42277.8 (from block UALVP42277) isolated scale. Hanke, 2001, Descr. p. 159

UALVP42277.9 (from block UALVP42277) isolated scale on matrix. Hanke, 2001, Descr. p. 159

UALVP42277.10 (from block UALVP42277) isolated scale. Hanke, 2001, Descr. p. 159

UALVP42277.11 (from block UALVP42277) isolated scale. Hanke, 2001, Descr. p. 159

UALVP42277.12 (from block UALVP42277) isolated scale. Hanke, 2001, Descr. p. 159

UALVP42277.13 (from block UALVP42277) two crushing plates fused together. Hanke, 2001, Descr. p. 159

UALVP42277.14 (from block UALVP42277) larger isolated crushing plate with a bit of another one fused to its side. Hanke, 2001, Descr. p. 159

UALVP42277.15 (from block UALVP42277) isolated crushing plate. Hanke, 2001, Descr. p. 159

UALVP42277.16 (from block UALVP42277) isolated crushing plate. Hanke, 2001, Descr. p. 159

UALVP42277.17 (from block UALVP42277) isolated, elongate crushing plate. Hanke, 2001, Descr. p. 159

UALVP42277.18 (from block UALVP42277) isolated crushing plate. Hanke, 2001, Descr. p. 159

UALVP42277.19 (from block UALVP42277) isolated crushing plate. Hanke, 2001, Descr. p. 159

UALVP42277.20 (from block UALVP42277) large, isolated crushing plate. Hanke, 2001, Descr. p. 159

UALVP42277.21 (from block UALVP42277) elongated isolated crushing plate—to be thin-sectioned. Hanke, 2001, Descr. p. 159

UALVP42277.22 (from block UALVP42277) elongated, isolated crushing plate. Hanke, 2001, Descr. p. 159

UALVP42277.23 (from block UALVP42277) isolated crushing plate. Hanke, 2001, Descr. p. 159

UALVP42277.24 (from block UALVP42277) slightly elongate, isolated crushing plate. Hanke, 2001, Descr. p. 159

UALVP42277.25 (from block UALVP42277) small, isolated crushing plate. Hanke, 2001, Descr. p. 159

UALVP42277.26 (from block UALVP42277) tiny broken tooth whorl—specimen lost (sand sized, must have popped off SEM stub) but still have SEM image information. Hanke, 2001, Descr. p. 159

UALVP42277.27 (from block UALVP42277) tiny whorl fragment—still on stub with UALVP42277.28. Hanke, 2001, Descr. p. 159

UALVP42277.28 (from block UALVP42277) tiny whorl fragment—on the SEM stub with UALVP42277.27. Hanke, 2001, Descr. p. 159

UALVP42277.29 (from block UALVP42277) nice tooth whorl—one we used to CT scan. Hanke, 2001, Descr. p. 159

UALVP42277.30 (from block UALVP42277) isolated tooth whorl. Hanke, 2001, Descr. p. 159

UALVP42277.31 (from block UALVP42277) isolated tooth whorl. Hanke, 2001, Descr. p. 159

UALVP42277.32 (from block UALVP42277) isolated tooth whorl—to be thin-sectioned. Hanke, 2001, p. 159

UALVP42277.33 (from block UALVP42277) isolated tooth whorl—to be thin-sectioned. Hanke, 2001, Descr. p. 159

UALVP42277.34 (from block UALVP42277) isolated tooth whorl. Hanke, 2001, Descr. p. 159

UALVP42277.35 (from block UALVP42277) isolated tooth whorl; smaller than the others, without the prominent row of larger, pointed cusps. Hanke, 2001, Descr. p. 159

UALVP42277.36 (from block UALVP42277) isolated tooth whorl. Hanke, 2001, Descr. p. 159

UALVP42277.37 (from block UALVP42277) isolated tooth whorl. Hanke, 2001, Descr. p. 159

UALVP42277.38 (from block UALVP42277) isolated tooth whorl and the newest cusp which broke off. Hanke, 2001, Descr. p. 159

UALVP42277.39 (from block UALVP42277) isolated cusp, broken off of a tooth whorl (none that it was associated with found). Hanke, 2001, Descr. p. 159

UALVP42277.40 (from block UALVP42277) isolated crushing plate; this specimen was thin-sectioned. Hanke, 2001, Descr. p. 159

UALVP44044 mostly head with mouth; has both whorls and crushing dentition. Hanke, 2001, Descr. p. 159, p. 434, Fig. 46, p. 169, Fig. 47. p. 171, Fig. 50.3, p. 182. CCIS L2-245, Devonian 14, 08

UALVP45204 isolated scale. Hanke, 2001, Descr. p. 159, fig. Fig. 49.1, p. 180

UALVP45205 isolated scale. Hanke, 2001, Descr. p. 159, fig. Fig. 49.2, p. 180

UALVP45206 isolated scale. Hanke, 2001, Descr. p. 159, fig. Fig. 49.3, p. 180

UALVP45207 isolated scale. Hanke, 2001, Descr. p. 159, fig. Fig. 49.4, p. 180

UALVP45208 isolated scale. Hanke, 2001, Descr. p. 159, fig. Fig. 49.5, p. 180

UALVP45209 isolated scale. Hanke, 2001, Descr. p. 159, fig. Fig. 49.6, p. 180

UALVP45210 isolated scale. Hanke, 2001, Descr. p. 159, fig. Fig. 49.7, p. 180

UALVP45211 isolated scale. Hanke, 2001, Descr. p. 159, fig. Fig. 50.1, p. 182

UALVP45212 isolated scale. Hanke, 2001, Descr. p. 159, fig. Fig. 50.2, p. 182

UALVP45308 isolated scale from 430.3 m. Hanke, 2001, Descr. p. 159, Fig. 146.1, p. 451

UALVP45309 isolated scale from 430.3 m. Hanke, 2001, Descr. p. 159, Fig. 146.2, p. 451

UALVP45310 isolated scale from 430.3 m. Hanke, 2001, Descr. p. 159, Fig. 146.3, p. 451

UALVP45311 isolated scale from 430.3 m. Hanke, 2001, Descr. p. 159, Fig. 146.4, p. 451

UALVP45312 isolated scale from 430.3 m. Hanke, 2001, Descr. p. 159, Fig. 146.5, p. 451

UALVP45313 isolated scale from 430.3 m. Hanke, 2001, Descr. p. 159, Fig. 146.6, p. 451

UALVP45314 isolated scale from 430.3 m. Hanke, 2001, Descr. p. 159, Fig. 146.7, p. 451

UALVP45315 isolated scale from 135.5 m. Hanke, 2001, Descr. p. 159, Fig. 146.8, p. 451

UALVP45316 isolated scale from 135.5 m. Hanke, 2001, Descr. p. 159, Fig. 146.9, p. 451

UALVP47187 scattered scales (**UALVP47187 =? UALVP32418**). Hanke, 2001, Descr. p. 159.? Not cited in Hanke, 2001, Ph.D, M.V.H. Wilson's copy of Hanke thesis has UALVP32418 crossed out, and written in pencil next to it is the catalogue number UALVP47187 L4-a, Teaching Collection, Biological Sciences Building Z-425, Unit 11, Drawer 1

Chondrichthyes Huxley, 1880
    Elasmobranchii Bonaparte, 1838
        Kathemacanthidae Gagnier and Wilson, 1996
UALVP32402 Holotype *Kathemacanthus rosulentus* **Gagnier and Wilson, 1996, p. 241**
Collector: Lindoe, L. Allan, 1990.
Canada: Northwest Territories, Mackenzie Mountains MOTH: Man on the Hill #1, Devonian-Devonian Early-Lochkovian,

Road River Fm. Fish, part and counterpart. Gagnier and Wilson (1996a), Hanke, 2001, Descr. p. 204, fig. Fig. 57, p. 206, Fig. 59.2, p. 210, Fig. 61.1, 61.6, p. 216, Fig. 62.2, p. 219. Hanke and Wilson, 2010, Descr. pp. 163–170 fig. Figs. 3–5, 6, A, 7–9. CCIS L2-245, Devonian 14, 04; CCIS L2-245, Devonian 14, 06

Gagnier, P.-Y., and Wilson, Mark V. H. 1996. Early Devonian acanthodians from northern Canada. Palaeontology 39(2):241–258.

Chondrichthyes Huxley, 1880
　Elasmobranchii Bonaparte, 1838
　　Altholepididae Hanke, 2001, p. 127
UALVP41485 Holotype *Altholepis spinata* Hanke, 2001, p. 145
Collector: Hanke, Gavin, 1996
Canada: Northwest Territories, Mackenzie Mountains, MOTH: Man on the Hill #1, Devonian-Devonian Early-Lochkovian, Road River Fm. Large-scale patch consisting of larger scales (identical to scales of UALVP41483) has two rows of intermediate spines and a large pectoral spine trailed by the pectoral fin. Hanke, 2001, Descr. p. 145, fig. Fig. 39, p. 148, Fig. 40.1–8, p. 152, Fig. 41.1–12, p. 154, Fig. 42.1–3, p. 157. CCIS L2-245, Devonian 14, 07

Hanke, Gavin F. 2001. Comparison of an Early Devonian Acanthodian and Putative Chondrichthyan assemblage using both isolated and articulated remains from the Mackenzie Mountains, with a cladistic analysis of Early Gnathostomes. Ph.D. thesis Department of Biological Sciences, University of Alberta, Edmonton, Alberta, Canada 566 pp.
PARATYPES:
UALVP45297 isolated scale. Hanke, 2001, Descr. p. 145, Fig. 145.1, p. 449
UALVP45298 isolated scale. Hanke, 2001, Descr. p. 145, Fig. 145.2, p. 449
UALVP45299 isolated scale. Hanke, 2001, Descr. p. 145, Fig. 145.3, p. 449

UALVP45300 isolated scale. Hanke, 2001, Descr. p. 145, Fig. 145.4, p. 449

UALVP45301 isolated scale. Hanke, 2001, Descr. p. 145, Fig. 145.5, p. 449

UALVP45302 isolated scale. Hanke, 2001, Descr. p. 145, Fig. 145.6, p. 449

UALVP45303 isolated scale. Hanke, 2001, Descr. p. 145, Fig. 145.7, p. 449

UALVP45304 isolated scale. Hanke, 2001, Descr. p. 145, Fig. 145.8, p. 449

UALVP45305 isolated scale. Hanke, 2001, Descr. p. 145, Fig. 145.9, p. 449

UALVP45306 isolated scale. Hanke, 2001, Descr. p. 145, Fig. 145.10, p. 449

UALVP45307 isolated scale. Hanke, 2001, Descr. p. 145, Fig. 145.11, p. 449

Chondrichthyes Huxley, 1880
    Elasmobranchii Bonaparte, 1838
        Altholepididae Hanke, 2001, p. 127
UALVP41498 Holotype *Altholepis davisi* **Hanke, 2001, p. 138**
Collector: Hanke, Gavin, 1996
Canada: Northwest Territories, Mackenzie Mountains, MOTH: Man on the Hill #1, Devonian-Devonian Early-Lochkovian, Road River Fm. Really long spine and a small patch of scales resembling *Cladolepis* spp., casts in teaching collection. Hanke, 2001, Descr. p. 138, Fig. 37.1–13, p. 140, Fig. 38.1–4, p. 142. CCIS L2-245, Devonian 14, 07

    Hanke, Gavin F. 2001. Comparison of an Early Devonian Acanthodian and Putative Chondrichthyan assemblage using both isolated and articulated remains from the Mackenzie Mountains, with a cladistic analysis of Early Gnathostomes. Ph.D. thesis Department of Biological Sciences, University of Alberta, Edmonton, Alberta, Canada 566 pp.

Chondrichthyes Huxley, 1880
Elasmobranchii Bonaparte, 1838
UALVP41488 HOLOTYPE *Obtusacanthus corroconis* Hanke and Wilson, 2004
Collector: Lindoe, L. Allan, 1996
Canada: Northwest Territories, Mackenzie Mountains MOTH: Man on the Hill #1 Devonian-Devonian Early-Lochkovian Road River Fm. Nearly intact body of possible chondrichthyan, all that is missing is the tail, on the 1996 "Wonder Block." Hanke, 2001, Descr. p. 60, Fig. 12, p. 62, Fig. 13, p. 64, Fig. 14.1–3, p. 66, Fig. 15.1–6, p. 69, Fig. 16.1–8, p. 72, Fig. 17.34, p. 75, Fig. 18.1, P. 78, Sahney and Wilson, 2001, JVP21(4):660–669 fig. Fig. 2, B, p. 663; Fig. 3, F, p. 664; Fig. 4, B, p. 666; Fig. 5, B, p. 667; Hanke and Wilson, 2004, pp. 189–215, Blais *et al.* 2011, JVP31(6):1189–1199 Fig. 7, A, B, p. 1198. CCIS L2-245, Devonian 14, 05

 Hanke, Gavin F., and Wilson, Mark V. H. 2004. New teleostome fishes and acanthodian systematics. pp. 189–215. *IN:* Arratia, G.; Wilson, Mark V. H.; Cloutier, R. (editors). Recent advances in the origin and early radiation of Vertebrates Dr. Friedrich Pfeil, München, Germany 703 pp.

PARATYPES:

UALVP19338 2 dorsal spines and preserved gut contents Type 3S scales. Hanke and Wilson 1997. Ichthyolith Issues Special Publication 2. IGCP 406 p. 15; Hanke, 2001, Descr. p. 60, Hanke and Wilson, 2004, Descr. pp. 189–215. CCIS L2-245, Devonian 14, 06
UALVP23349 body parts with partial tail Type 3S scales. Hanke and Wilson 1997. Ichthyolith Issues Special Publication 2. IGCP 406 Descr. p. 15, Hanke, 2001, Descr. p. 60, Fig. 17.29–33, p. 75, Fig. 18.4–8, p. 78, Hanke and Wilson, 2004, Descr. p. 196, Fig. 9 (l–n), 10 a–c. CCIS L2-245, Devonian 05, 01
UALVP41503 patch of 3S scales including head & flank scales & a dorsal? Spine partly covered by *Lepidaspis* sp. Hanke, 2001, Descr. p. 60, Fig. 17.1–24, 17.28, p. 75, Hanke and Wilson, 2004, Descr. pp. 189–215. CCIS L2-245, Devonian 14, 06

UALVP41764 Hanke, 2001, Descr. p. 60. Hanke and Wilson, 2004, Descr. pp. 189–215. CCIS L2-245, Devonian 14, 06
UALVP43939 patch of scales. Hanke, 2001, p. 60, Hanke and Wilson, 2004, Descr. pp. 189–215. CCIS L2-245, Devonian 14, 06
UALVP43942 patch of scales. Hanke and Wilson, 2004, Descr. p. 196, Hanke, 2001, Descr. p. 60, Hanke and Wilson, 2004, Descr. pp. 189–215. CCIS L2-245, Devonian 14, 06
UALVP43943 patch of scales. Hanke, 2001, Descr. p. 60, Hanke and Wilson, 2004, Descr. pp. 189–215. CCIS L2-245, Devonian 14, 06
UALVP43945 Hanke, 2001, Descr. p. 60, Hanke and Wilson, 2004, Descr. pp. 189–215. CCIS L2-245, Devonian 14, 09
UALVP45286 isolated scale. Hanke, 2001, Descr. p. 60, Fig. 144.1, p. 447, Hanke and Wilson, 2004, Descr. pp. 189–215
UALVP45287 isolated scale. Hanke, 2001, Descr. p. 60, Fig. 144.2, p. 447, Hanke and Wilson, 2004, Descr. pp. 189–215
UALVP45288 isolated scale. Hanke, 2001, Descr. p. 60, Fig. 144.3, p. 447, Hanke and Wilson, 2004, Descr. pp. 189–215
UALVP45289 isolated scale. Hanke, 2001, Descr. p. 60, Fig. 144.4, p. 447, Hanke and Wilson, 2004, Descr. pp. 189–215
UALVP45290 isolated scale. Hanke, 2001, Descr. p. 60, Fig. 144.5, p. 447, Hanke and Wilson, 2004, Descr. pp. 189–215
UALVP45291 isolated scale. Hanke, 2001, Descr. p. 60, Fig. 144.6, p. 447, Hanke and Wilson, 2004, Descr. pp. 189–215
UALVP45292 isolated scale. Hanke, 2001, Descr. p. 60, Fig. 144.7, p. 447, Hanke and Wilson, 2004, Descr. pp. 189–215
UALVP45293 isolated scale. Hanke, 2001, Descr. p. 60, Fig. 144.8, p. 447, Hanke and Wilson, 2004, Descr. pp. 189–215
UALVP45294 isolated scale. Hanke, 2001, Descr. p. 60, Fig. 144.9, p. 447, Hanke and Wilson, 2004, Descr. pp. 189–215

Chondrichthyes Huxley, 1880
    Elasmobranchii Bonaparte, 1838
        Elegestolepidida Andreev, Coates, Karajutè-Talimaa, Shelton, Cooper and Sansom, 2017

Elegestolepididae Andreev, Coates, Karajutè-Talimaa, Shelton, Cooper and Sansom, 2017
UALVP41489 HOLOTYPE *Platylepis cummingi* **Hanke, 2001, Descr. p. 95**
Collector: Lindoe, L. Allan, 1996
Canada: Northwest Territories, Mackenzie Mountains, MOTH: Man on the Hill #1, Devonian-Devonian Early-Lochkovian, Road River Fm. A large patch of scales resembling *Ellesmereia* (Vieth 1980), Hanke, 2001, Descr. p. 95, fig. Fig. 24, p. 98, Fig. 25, 1–6, p. 101, Fig. 26, 1–18, p. 103, Fig. 27, 1 4, p. 105

Hanke, Gavin F. 2001. Comparison of an Early Devonian Acanthodian and Putative Chondrichthyan assemblage using both isolated and articulated remains from the Mackenzie Mountains, with a cladistic analysis of Early Gnathostomes. Ph.D. thesis Department of Biological Sciences, University of Alberta, Edmonton, Alberta, Canada 566 pp.

PARATYPES:
UALVP45331 isolated scale, a large patch of scales resembling *Ellesmereia* (Vieth, 1980), from 430.3 meters. Hanke, 2001, Descr. 95, & fig. Fig. 148.1, p. 455
UALVP45332 isolated scale from 430.3 m. Hanke, 2001, Descr. p. 95, & fig. Fig. 148.2, p. 455
UALVP45333 isolated scale from 430.3 m. Hanke, 2001, Descr. p. 95, & fig. Fig. 148.3, p. 455
UALVP45334 isolated scale from 430.3 m. Hanke, 2001, Descr. p. 95, & fig. Fig. 148.4, p. 455
UALVP45335 isolated scale from 430.3 m. Hanke, 2001, Descr. p. 95, & fig. Fig. 148.5, p. 455
UALVP45336 isolated scale from 430.3 m. Hanke, 2001, Descr. p. 95, & fig. Fig. 148.6, p. 455
UALVP45337 isolated scale from 430.3 m. Hanke, 2001, Descr. p. 95, & fig. Fig. 148.7, p. 455
UALVP45338 isolated scale from 430.3 m. Hanke, 2001, Descr. p. 95, & fig. Fig. 148.8, p. 455

UALVP45339 isolated scale from 430.3 m. Hanke, 2001, Descr. p. 95, & fig. Fig. 148.9, p. 455

UALVP45340 isolated scale from 430.3 m. Hanke, 2001, Descr. p. 95, & fig. Fig. 148.10, p. 455

UALVP45341 isolated scale from 135.3 m. Hanke, 2001, Descr. p. 95, & fig. Fig. 148.11, p. 455

UALVP45342 isolated scale from 135.5 m. Hanke, 2001, Descr. p. 95, & fig. Fig. 148.12, p. 455

UALVP45343 isolated scale from 135.5 m. Hanke, 2001, Descr. p. 95, & fig. Fig. 148.13, p. 455

UALVP45344 isolated scale from 135.5 m. Hanke, 2001, Descr. p. 95, & fig. Fig. 148.14, p. 455

Chondrichthyes Huxley, 1880
   Elasmobranchii Bonaparte, 1838
      Elegestolepidida Andreev, Coates, Karajutè-Talimaa, Shelton, Cooper and Sansom, 2017
         Kannathalepididae Märss and Gagnier, 2001

UALVP44886 HOLOTYPE *Kannathalepis milleri* **Märss and Gagnier, 2001, p. 695**

Collectors: Wilson, Mark V. H., Gagnier, Pierre-Yves, and Märss, Tiiu, 1994. Canada: Nunavut, Baillie-Hamilton Island, BH-2–94 28.5 m, Silurian-Silurian Early-Wenlock-Wenlock Late-Homerian Early, Cape Phillips Fm., shale. Thin section of scale. Marss and Gagnier, 2001, Descr. p. 695, p. 696, Fig. 3, A, D. CCIS L2-245, Silurian 11, 07

Märss, Tiiu, and Gagnier, P.-Y. 2001. A new chondrichthyan from the Wenlock, Lower Silurian, of Baillie-Hamilton Island, the Canadian Arctic. Journal of Vertebrate Paleontology. Vol. 21(4):693–701.

PARATYPES:

UALVP44605 scale. Märss and Gagnier, 2001, Descr. p. 695. p. 697, Fig. 4, F

UALVP44607 scale. Märss and Gagnier. 2001, Descr. p. 695. p. 697, Fig. 4, D

UALVP44608 scale. Märss and Gagnier. 2001, Descr. p. 695, & fig. Fig. 3, G–I., p. 696

UALVP44609 scale. Märss and Gagnier. 2001, Descr. p. 695. & fig. p. 696, Fig. 3, F.

UALVP44610 scale mounted on STUB118. Märss and Gagnier, 2001, Descr. p. 695, p. 696, Fig. 3, C.

UALVP44611 scale. Märss and Gagnier, 2001, Descr. p. 695, p. 697, Fig. 4, B. STUB118

UALVP44612 scale. Märss and Gagnier. 2001, Descr. p. 695, p. 697, Fig. 4, C. STUB118

UALVP44882 scale mounted on STUB129. Märss and Gagnier, 2001, Descr. p. 695, Fig. 3, J, p. 696.

UALVP44883 scale mounted on STUB129. Märss and Gagnier, 2001, Descr. p. 695, Fig. 4, A, p. 697

UALVP44884 scale mounted on STUB129. Märss and Gagnier, 2001, Descr. p. 695, p. 697, Fig. 4, H

UALVP44885 scale mounted on STUB129. Märss and Gagnier, 2001, Descr. p. 695, p. 697, Fig. 4, G

UALVP44886 scale mounted on STUB129. Märss and Gagnier, 2001, Descr. p. 695, p. 696, Fig. 3, A, D

UALVP44887 scale mounted on STUB129. Märss and Gagnier, 2001, Descr. p. 695, p. 696, Fig. 3, B, E

UALVP44889 scale mounted on STUB129. Märss and Gagnier, 2001, Descr. p. 695, p. 697, Fig. 4, E

UALVP45043 thin section of scale. Märss and Gagnier, 2001, Descr. p. 695, pp. 698–699, Fig. 5, D, p. 698. CCIS L2-245, Silurian 11, 07

UALVP45044 thin section of scale. Märss and Gagnier, 2001, Descr. p. 695, pp. 698–699, Fig. 5, E. CCIS L2-245, Silurian 11, 07

UALVP45045 thin section of scale. Märss and Gagnier, 2001, Descr. p. 695, pp. 698–699, Fig. 5, F. CCIS L2-245, Silurian 11, 07

UALVP45046 thin section of scale. Märss and Gagnier, 2001, Descr. p. 695, pp. 698–699, Fig. 5, G

UALVP45047 scale photographed in anise oil. Märss and Gagnier, 2001, Descr. p. 695, pp. 698–699, Fig. 5, A

UALVP45048 scale photographed in anise oil. Märss and Gagnier, 2001, Descr. p. 695, pp. 698–699, Fig. 5, B

UALVP45049 scale photographed in anise oil. Märss and Gagnier. 2001. Descr. p. 695. & fig. Fig. 5, C, p. 698

UALVP45050 scale photographed in anise oil. Märss and Gagnier. 2001. Descr. p. 695. pp. 698–699, Fig. 5, H

Chondrichthyes Huxley, 1880
Elasmobranchii Bonaparte, 1838
UALVP42180 HOLOTYPE *Arrapholepis valyalamia* **Hanke, 2001, p. 83**
Collector: Lindoe, L. Allan, 1996
Canada: Northwest Territories, Mackenzie Mountains, MOTH: Man on the Hill #1, Devonian-Devonian Early-Lochkovian, Road River Fm. Hanke, 2001, Descr. p. 83, Fig. 21, 3, p. 87. CCIS L2-245, Devonian 14, 08

Hanke, Gavin F. 2001. Comparison of an Early Devonian Acanthodian and Putative Chondrichthyan assemblage using both isolated and articulated remains from the Mackenzie Mountains, with a cladistic analysis of Early Gnathostomes. Ph.D. thesis Department of Biological Sciences, University of Alberta, Edmonton, Alberta, Canada 566 pp.

PARATYPES:
UALVP41520 Hanke, 2001, Descr. p. 83, Fig. 20.4–6, p. 85, Fig. 21, 6, p. 87. CCIS L2-245, Devonian 14, 08

UALVP41562 Hanke, 2001, Descr. p. 83, Fig. 20.2, 20.3, p. 85, Fig. 21, 1, p. 87. CCIS L2-245, Devonian 14, 08

UALVP41687 Hanke, 2001, Descr. p. 83. CCIS L2-245, Devonian 14, 08

UALVP41776 Hanke, 2001, Descr. p. 83. CCIS L2-245, Devonian 14, 08

UALVP41799 Hanke, 2001, Descr. p. 83. CCIS L2-245, Devonian 14, 08

UALVP41980 Hanke, 2001, Descr. p. 83. CCIS L2-245, Devonian 14, 08

UALVP41995 Hanke, 2001, Descr. p. 83. CCIS L2-245, Devonian 14, 08

UALVP41999 Hanke, 2001, Descr. p. 83. CCIS L2-245, Devonian 14, 08

UALVP43944 Hanke, 2001, Descr. p. 83. CCIS L2-245, Devonian 14, 08

UALVP43947 Hanke, 2001, Descr. p. 83

UALVP45161 isolated scale. Hanke, 2001, Descr. p. 83, Fig. 22.1, p. 89

UALVP45162 isolated scale. Hanke, 2001, Descr. p. 83, Fig. 22.2, p. 89

UALVP45163 isolated scale. Hanke, 2001, Descr. p. 83, Fig. 22.3, p. 89

UALVP45164 isolated scale. Hanke, 2001, Descr. p. 83, Fig. 22.4, p. 89

UALVP45165 isolated scale. Hanke, 2001, Descr. p. 83, Fig. 22.5, p. 89

UALVP45166 isolated scale. Hanke, 2001, Descr. p. 83, Fig. 22.6, p. 89

UALVP45167 isolated scale. Hanke, 2001, Descr. p. 83, Fig. 22.7, p. 89

UALVP45168 isolated scale. Hanke, 2001, Descr. p. 83, Fig. 22.8, p. 89

UALVP45169 isolated scale. Hanke, 2001, Descr. p. 83, Fig. 22.9, p. 89

UALVP45170 isolated scale. Hanke, 2001, Descr. p. 83, Fig. 22.10, p. 89

UALVP45171 isolated scale. Hanke, 2001, Descr. p. 83, Fig. 22.11, p. 89

UALVP45172 isolated scale. Hanke, 2001, Descr. p. 83, Fig. 22.12, p. 89

UALVP45173 isolated scale. Hanke, 2001, Descr. p. 83, Fig. 22.13, p. 89

UALVP45317 isolated scale from 430.3 m. Hanke, 2001, Descr. p. 83 Fig. 147.1, p. 453

UALVP45318 isolated scale from 430.3 m. Hanke, 2001, Descr. p. 83, Fig. 147.2, p. 453

UALVP45319 isolated scale from 430.3 m. Hanke, 2001, Descr. p. 83, Fig. 147.3, p. 453

UALVP45320 isolated scale from 430.3 m. Hanke, 2001, Descr. p. 83, Fig. 147.4, p. 453

UALVP45321 isolated scale from 430.3 m. Hanke, 2001, Descr. p. 83, Fig. 147.5, p. 453

UALVP45322 isolated scale from 430.3 m. Hanke, 2001, Descr. p. 83, Fig. 147.6, p. 453

UALVP45323 isolated scale from 430.3 m. Hanke, 2001, Descr. p. 83, Fig. 147.7, p. 453

UALVP45324 isolated scale from 430.3 m. Hanke, 2001, Descr. p. 83, Fig. 147.8, p. 453

UALVP45325 isolated scale from 430.3 m. Hanke, 2001, Descr. p. 83, Fig. 147.9, p. 453

UALVP45326 isolated scale from 430.3 m. Hanke, 2001, Descr. p. 83, Fig. 147.10, p. 453

UALVP45327 isolated scale from 430.3 m. Hanke, 2001, Descr. p. 83, Fig. 147.11, p. 453

UALVP45328 isolated scale from 430.3 m. Hanke, 2001, Descr. p. 83, Fig. 147.12, p. 453

UALVP45329 isolated scale from 430.3 m. Hanke, 2001, Descr. p. 83, Fig. 147.13, p. 453

UALVP45330 isolated scale from 135.5 m. Hanke, 2001, Descr. p. 83, Fig. 147.14, p. 453

Chondrichthyes Huxley, 1880
  Elasmobranchii Bonaparte, 1838
UALVP46531 HOLOTYPE *Wapitiodus homalorhizo* **Mutter, De Blanger, and Neuman, 2007**
Collector: Lindoe, Allan L., 2003.
Canada: British Columbia, Wapiti Lake, Wapiti Lake F. Triassic-Triassic Lower-Smithian, Sulphur Mountain Fm. F cirque, scree

slope on E side of ridge, W of Fossil F. Lake. Nearly complete, two parts. Mutter, De Blanger, and Neuman, 2007, fig. Fig. 13, A, B, p. 323, Fig. 14, p. 324, Fig. 15, p. 326; Mutter, R. 2004. Fossil fische aus der Trias der kanadischen Rocky Mountians. Described this specimen as *Palaeobates*, Hybodontidae

Mutter, R. J., De Blanger, K., and Neuman, A. G. 2007. Elasmobranchs from the Lower Triassic Sulphur Mountain Formation near Wapiti Lake (BC, Canada). Zoological Journal of the Linnean Society Vol. 149:309–337.

Chondrichthyes Huxley, 1880
    Holocephali Bonaparte 1832–1841
        Bradyodontida Woodward, 1921
            Chimaerina Zangerl, 1981
                Cochliodontidae Owen, 1867
UALVP47002 HOLOTYPE *Listracanthus pectenatus* **Mutter and Neuman, 2006, p. 273**
Collector: McClafferty, Phyllis, 2004.
Canada: British Columbia, Wapiti Lake, Wapiti Lake C. Triassic-Triassic Lower-Smithian, Sulphur Mountain Fm. Shagreen, cluster of denticles. Mutter and Neuman, 2006, Descr. p. 273, & fig. Fig. 2, A–C, p. 274. CCIS L2-245, Triassic 11, 04

Mutter, Raoul J., and Neuman, Andrew G. 2006. An enigmatic chondrichthyan with Paleozoic affinities from the Lower Triassic of western Canada. Acta Palaeonologica Polonica Vol. 52(2):271–282

PARATYPES:
UALVP1840 (=? Incorrectly cited catalogue number in Mutter and Neuman, 2006, Descr. p. 273. Catalogue says this specimen is a tooth from the Mississippian of Alberta)
UALVP1843 (=? Incorrectly cited catalogue number in Mutter and Neuman, 2006, Descr. p. 273. Catalogue says this specimen is a tooth plate from the Mississippian of Alberta)

UALVP1885 1 scale. Mutter and Neuman, 2006, Descr. p. 273. CCIS L2-245, Triassic 02, 08

UALVP1886 1 scale A/B. Mutter and Neuman, 2006, Descr. p. 273. CCIS L2-245, Triassic 02, 08

UALVP1887 1 scale A/B. Mutter and Neuman, 2006, Descr. p. 273. CCIS L2-245, Triassic 02, 08

UALVP1888 1 scale A/B. Mutter and Neuman, 2006, Descr. p. 273. CCIS L2-245, Triassic 02, 08

UALVP1889 dermal denticle. Mutter and Neuman, 2006, Descr. p. 274, fig. Fig. 7A, p. 279.

UALVP1891 1 scale A/B. Mutter and Neuman, 2006, Descr. p. 273. CCIS L2-245, Triassic 02, 08

UALVP1892 1 scale A/B. Mutter and Neuman, 2006, Descr. p. 273. CCIS L2-245, Triassic 02, 08

UALVP1893 1 scale A/B. Mutter and Neuman, 2006, Descr. p. 273. CCIS L2-245, Triassic 02, 08

UALVP1894 1 scale A/B. Mutter and Neuman, 2006, Descr. p. 273. CCIS L2-245, Triassic 02, 08

UALVP1895 1 scale A/B. Mutter and Neuman, 2006, Descr. p. 273. CCIS L2-245, Triassic 02, 08

UALVP1896 1 scale A/B. Mutter and Neuman, 2006, Descr. p. 273. CCIS L2-245, Triassic 02, 08

UALVP1899 1 scale. Mutter and Neuman, 2006, Descr. p. 273. CCIS L2-245, Triassic 02, 08

UALVP17931 scales. Mutter and Neuman, 2006, Descr. p. 273, Mutter, De Blanger, and Neuman, 2007, Zool. J. of the Linn. Soc., Vol. 149, p. 329

UALVP17938 placoid scales. Mutter and Neuman, 2006, Descr. p. 273, p. 278, fig. Fig. 3, A, p. 275, Mutter and Neuman, 2008, Paleo., Paleo., Paleo. CCIS L2-245, Triassic 04, 13

UALVP17940 fin spines. Mutter and Neuman, 2006, Descr. p. 273, Mutter and Neuman, 2008, Paleo., Paleo., Paleo.

UALVP38562 spines. Thin sections 8, 9, 10. Mutter and Neuman, 2006, Descr. p. 274, p. 278, fig. Fig. 6, A and B, p. 277, Fig. 7, B, p. 279. CCIS L2-245, Triassic 05, 08

UALVP46540 Shagreen. Mutter and Neuman, 2006, Descr. p. 273, p. 278, fig. Fig. 3, B_1, B_2, B_3, p. 275. Mutter and Neuman, 2008, Paleo., Paleo., Paleo. CCIS L2-245, Triassic 04, 13

UALVP46541 spine. Mutter and Neuman, 2006, Descr. p. 273, Mutter and Neuman, 2008, Paleo., Paleo., Paleo. CCIS L2-245, Triassic 04, 13

UALVP46542 spines, three parts "L." Mutter and Neuman, 2006, Descr. p. 273, Mutter and Neuman, 2008, Paleo., Paleo., Paleo. CCIS L2-245, Triassic 04, 13

UALVP46543 spine. Mutter and Neuman, 2006, Acta Paleont. Pol, Descr. p. 273. CCIS L2-245, Triassic 04, 13

UALVP46544 spine. Mutter and Neuman, 2006, Descr. p. 273, Mutter and Neuman, 2008, Paleo., Paleo., Paleo. CCIS L2-245, Triassic 04, 13

UALVP46545 spine. Mutter and Neuman, 2006, Descr. p. 273, Mutter and Neuman, 2008, Paleo., Paleo., Paleo. CCIS L2-245, Triassic 04, 13

UALVP46546 spine. Mutter and Neuman, 2006, Descr. p. 273, Mutter and Neuman, 2008, Paleo., Paleo., Paleo. CCIS L2-245, Triassic 04, 13

UALVP46547 spine. Mutter and Neuman, 2006, Descr. p. 273, Mutter and Neuman, 2008, Paleo., Paleo., Paleo. CCIS L2-245, Triassic 04, 13

UALVP46549 spine. Mutter and Neuman, 2006, Descr. p. 273, Mutter and Neuman, 2008, Paleo., Paleo., Paleo. CCIS L2-245, Triassic 04, 13

UALVP46550 spine. Mutter and Neuman, 2006, Descr. p. 273, Mutter and Neuman, 2008, Paleo., Paleo., Paleo. CCIS L2-245, Triassic 04, 13

UALVP46551 spine. Mutter and Neuman, 2006, Descr. p. 273, fig. Fig. 4, A–D, p. 276, Mutter and Neuman, 2008, Paleo., Paleo., Paleo. CCIS L2-245, Triassic 04, 13

UALVP46552 spine. Mutter and Neuman, 2006, Descr. p. 273, Mutter and Neuman, 2008, Paleo., Paleo., Paleo. CCIS L2-245, Triassic 04, 13

UALVP46553 spine. Mutter and Neuman, 2006, Descr. p. 273, Mutter and Neuman, 2008, Paleo., Paleo., Paleo. CCIS L2-245, Triassic 04, 13

UALVP46554 spine. Mutter and Neuman, 2006, Descr. p. 273, Mutter and Neuman, 2008, Paleo., Paleo., Paleo. CCIS L2-245, Triassic 04, 13

UALVP46555 spine +? two parts. Mutter and Neuman, 2006, Descr. p. 273, Mutter and Neuman, 2008, Paleo., Paleo., Paleo.

UALVP46556 spine +? two parts. Mutter and Neuman, 2006, Descr. p. 273, Mutter and Neuman, 2008, Paleo., Paleo., Paleo.

UALVP46557 spine, two parts. Mutter and Neuman, 2006, Descr. p. 273, Mutter and Neuman, 2008, Paleo., Paleo., Paleo.

UALVP46558 spine. Mutter and Neuman, 2006, Descr. p. 273, Mutter and Neuman, 2008, Paleo., Paleo., Paleo.

UALVP46559 spine. Mutter and Neuman, 2006, Descr. p. 273, Mutter and Neuman, 2008, Paleo., Paleo., Paleo.

UALVP46560 spine. Mutter and Neuman, 2006, Descr. p. 273, Mutter and Neuman, 2008, Paleo., Paleo., Paleo. CCIS L2-245, Triassic 05, 09

UALVP46561 spine. Mutter and Neuman, 2006, Descr. p. 273, Mutter and Neuman, 2008, Paleo., Paleo., Paleo. CCIS L2-245, Triassic 05, 09

UALVP46562 spine. Mutter and Neuman, 2006, Descr. p. 273, Mutter and Neuman, 2008, Paleo., Paleo., Paleo. CCIS L2-245, Triassic 05, 09

UALVP46563 spine. Mutter and Neuman, 2006, Descr. p. 273, Mutter and Neuman, 2008, Paleo., Paleo., Paleo. CCIS L2-245, Triassic 05, 09

UALVP46564 spine. Mutter and Neuman, 2006, Descr. p. 273, Mutter and Neuman, 2008, Paleo., Paleo., Paleo. CCIS L2-245, Triassic 05, 09

UALVP46565 spine. Mutter and Neuman, 2006, Descr. p. 273, Mutter and Neuman, 2008, Paleo., Paleo., Paleo. CCIS L2-245, Triassic 05, 09

UALVP46566 spine. Mutter and Neuman, 2006, Descr. p. 273, Mutter and Neuman, 2008, Paleo., Paleo., Paleo. CCIS L2-245, Triassic 05, 09

UALVP46567 spine, two parts. Mutter and Neuman, 2006, Descr. p. 273, Mutter and Neuman, 2008, Paleo., Paleo., Paleo.

UALVP46568 Shagreen, Mutter and Neuman, 2006, Descr. p. 273, p. 276, p. 278, Mutter and Neuman, 2008, Paleo., Paleo., Paleo. CCIS L2-245, Triassic 05, 08

UALVP46569 spine. Mutter and Neuman, 2006, Descr. p. 273, Mutter, 2004, Vierteljahrsschrift der Naturforschenden Gesellschaft in Zürich, Vol. 149:51–58, Mutter and Neuman, 2008, Paleo., Paleo., Paleo. CCIS L2-245, Triassic 05, 08

UALVP46570 spine. Mutter and Neuman, 2006, Descr. p. 273, Mutter and Neuman, 2008, Paleo., Paleo., Paleo., Mutter, 2004, Vierteljahrsschrift der Naturforschenden Gesellschaft in Zürich, Vol. 149:51–58. CCIS L2-245, Triassic 05, 08

UALVP46571 spine. Mutter and Neuman, 2006, Descr. p. 273, Mutter and Neuman, 2008, Paleo., Paleo., Paleo. CCIS L2-245, Triassic 05, 08

UALVP46573 Shagreen, fin spines. Mutter and Neuman, 2006, Descr. p. 273, p. 278, + THIN SECTION UALVP46573-T1

UALVP46574 spine. Mutter and Neuman, 2006, Descr. p. 273, Mutter and Neuman, 2008, Paleo., Paleo., Paleo. CCIS L2-245, Triassic 05, 08

UALVP46575 spine. Mutter and Neuman, 2006, Descr. p. 273, p. 275. Mutter and Neuman, 2008, Paleo., Paleo., Paleo. CCIS L2-245, Triassic 05, 08

UALVP46576 spine, two parts. Mutter and Neuman, 2006, Descr. p. 273. Mutter and Neuman, 2008, Paleo., Paleo., Paleo. CCIS L2-245, Triassic 05, 08

UALVP46577 spine. Mutter and Neuman, 2006, Descr. p. 273, p. 275. Mutter and Neuman, 2008, Paleo., Paleo., Paleo. CCIS L2-245, Triassic 05, 08

UALVP46578 spine, small, two parts. Talus slope. Mutter and Neuman, 2006, Descr. p. 273. Mutter and Neuman, 2008, Paleo., Paleo., Paleo. CCIS L2-245, Triassic 05, 08

UALVP46792 spine. Mutter and Neuman, 2006, Descr. p. 273. Mutter and Neuman, 2008, Paleo., Paleo., Paleo. CCIS L2-245, Triassic 05, 14

UALVP47001 isolated denticles plus isolated tooth (polyacrodontoid tooth). Mutter and Neuman, 2006, Descr. p. 273. Mutter and Neuman, 2008, Paleo., Paleo., Paleo. CCIS L2-245, Triassic 11, 04

UALVP47003 cluster of denticles. Mutter and Neuman, 2006, Descr. p. 273, Fragment of tooth whorl & adjoining lateral tooth files. Tooth whorl for *Caseodus varidentis* described in Mutter and Neuman, 2008, Geol. Soc. Special Pub. No. 295: 9–41. Descr. pp. 17–18, & fig. Fig. 8, a, b, p. 17. CCIS L2-245, Triassic 11, 04

UALVP47004 cluster of denticles. Mutter and Neuman, 2006, Descr. p. 273, p. 276, p. 278. Mutter and Neuman, 2008, Paleo., Paleo., Paleo. CCIS L2-245, Triassic 11, 04

UALVP47005 isolated denticles plus phosphatic structure. Mutter and Neuman, 2006, Descr. p. 273, p. 276, p. 278. Mutter and Neuman, 2008, Paleo., Paleo., Paleo. CCIS L2-245, Triassic 11, 04

UALVP47006 isolated denticle. Mutter and Neuman, 2006, Descr. p. 273. Mutter and Neuman, 2008, Paleo., Paleo., Paleo. CCIS L2-245, Triassic 11, 04

UALVP47007 isolated denticle. Mutter and Neuman, 2006. Descr. p. 273. Mutter and Neuman, 2008, Paleo., Paleo., Paleo. CCIS L2-245, Triassic 11, 04

UALVP47008 isolated denticle. Mutter and Neuman, 2006. Descr. p. 273. Mutter and Neuman, 2008, Paleo., Paleo., Paleo. CCIS L2-245, Triassic 11, 04

UALVP47009 isolated denticle. Mutter and Neuman, 2006. Descr. p. 273. Mutter and Neuman, 2008, Paleo., Paleo., Paleo. CCIS L2-245, Triassic 11, 04

UALVP47010 isolated denticle. Mutter and Neuman, 2006. Descr. p. 273. Mutter and Neuman, 2008, Paleo., Paleo., Paleo. CCIS L2-245, Triassic 11, 03

UALVP47011 isolated large denticle Mutter and Neuman, 2006. Descr. p. 273. Mutter and Neuman, 2008, Paleo., Paleo., Paleo. CCIS L2-245, Triassic 11, 03

UALVP47012 isolated denticle plus shrimps. Mutter and Neuman, 2006. Descr. p. 273. Mutter and Neuman, 2008, Recovery from the end-Permian extinction event: evidence from "Lilliput *Listracanthus*," Paleo., Paleo., Paleo. CCIS L2-245, Triassic 11, 03
UALVP47013 isolated denticles. Mutter and Neuman, 2006. Descr. p. 273. Mutter and Neuman, 2008, Recovery from the end-Permian extinction event: evidence from "Lilliput *Listracanthus*," Paleo., Paleo., Paleo. CCIS L2-245, Triassic 11, 03
UALVP47014 cluster of small denticles. Mutter and Neuman, 2006. Descr. p. 273, p. 278. Mutter and Neuman, 2008, Recovery from the end-Permian extinction event: evidence from "Lilliput *Listracanthus*," Paleo., Paleo., Paleo. CCIS L2-245, Triassic 11, 03
UALVP47015 cluster of denticles Mutter and Neuman, 2006. Descr. p. 273, p. 276, p. 278. Mutter and Neuman, 2008, Recovery from the end-Permian extinction event: evidence from "Lilliput *Listracanthus*," Paleo., Paleo., Paleo. CCIS L2-245, Triassic 11, 03

Chondrichthyes Huxley, 1880
    Elasmobranchii Bonaparte, 1838
        Lamniformes Berg, 1958
            Odontaspididae Müller and Henle, 1839
UALVP53199 HOLOTYPE *Odontaspis watinensis* **Cook, Wilson, Murray, Plint, Newbrey, and Everhart, 2013, p. 572**
Collectors: Lindoe, L. Allan, Neuman, Andrew G., 1983
Canada: Alberta, Watino. Watino Outcrop. Out crop at beginning of curve to N, on S bank of Smoky River downstream from Watino Bridge. 81–4. Cretaceous-Cretaceous Upper-early Turonian, Kaskapau Fm. Anterior tooth. Cook *et al.*, 2013, Descr. p. 572, Fig. 11, F, p. 571. CCIS L2-245, Cretaceous 04, 15

    Cook, Todd D., Wilson, Mark V. H., Murray, Alison M., Plint, A. Guy, Newbrey, Michael G., and Everhart, Michael J. 2013. A high latitude euselachian assemblage from the early Turonian of Alberta, Canada. Journal of Systematic Palaeontology Vol. 11(5):555–587.

PARATYPES:

UALVP53200 lateral tooth. Cook *et al.*, 2013, Descr. p. 572, Fig. 11, G, p. 571. CCIS L2-245, Cretaceous 04, 15

UALVP53201 lateral tooth. Cook *et al.*, 2013, Descr. p. 572, Fig. 11, H, p. 571. CCIS L2-245, Cretaceous 04, 15

UALVP53202 lateral tooth. Cook *et al.*, 2013, Descr. p. 572, Fig. 11, I, p. 571. CCIS L2-245, Cretaceous 04, 15

Chondrichthyes Huxley, 1880
 Elasmobranchii Bonaparte, 1838
  Batomorphii Cappetta, 1980
   Myliobatiformes Compagno, 1973
    Rhinopteridae Jordan and Evermann, 1896
UALVP52360 HOLOTYPE *Eorhinoptera grabdai* **Case, Cook, and Wilson, 2011**

Collector: Grabda, David W.

United States: South Carolina Fishburne Tertiary-Eocene-Eocene early (Ypresian) single isolated pavement tooth, with 12 root lobelets. Case *et al.*, 2010, Descr. pp. 1–6, p. 2, Fig. 2, A, p. 3, Fig. 3, A, p. 4. Case *et al.*, 2011, fig. Fig. 2, A, p. 141, Fig. 3, A, p. 142. CCIS L2-245, Eocene 04, 12

 Case, Gerard R., Cook, Todd D., and Wilson, Mark V. H. 2011. A new genus and species of fossil myliobatoid ray from the Fishburne Formation (lower Eocene/Ypresian) of Berkeley County, South Carolina, USA. Historical Biology, Vol. 23 (2–3):139–144.

PARATYPES:

UALVP52361 single isolated pavement tooth, with four root lobelets. Case *et al.*, 2010, Descr. pp. 1–6, p. 2, fig. Fig. 2, B, p. 3, Fig. 3, B, p. 4, Case *et al.*, 2011, fig. Fig. 2, B, p. 141, Fig. 3, B, p. 142. CCIS L2-245, Eocene 04, 12

UALVP52362 single isolated pavement tooth, with six root lobelets. Case *et al.*, 2011, fig. Fig. 2, C, p. 141. CCIS L2-245, Eocene 04, 12

UALVP52363 single isolated pavement tooth, with four root lobelets. Case *et al.*, 2010, Descr. pp. 1–6, p. 2, Case *et al.*, 2011, Descr. p. 140. CCIS L2-245, Eocene 04, 12

UALVP52364 under ten, fractured pavement teeth. Case *et al.*, 2010, Descr. pp. 1–6, p. 2, fig. Fig. 2, D, p. 3, Case *et al.*, 2011, fig. Fig. 2, D, p. 141. CCIS L2-245, Eocene 04, 12

Chondrichthyes Huxley, 1880
　　Elasmobranchii Bonaparte, 1838
　　　　Batomorphii Cappetta, 1980
　　　　　　Rajiformes Berg, 1940
　　　　　　　　Sclerorhynchoidei Cappetta, 1980
　　　　　　　　　Sclerorhynchidae Cappetta, 1974
UALVP53724 HOLOTYPE *Borodinopristis shannoni* Case, Cook, Wilson, and Borodin, 2012, p. 593
Collector: Case, Gerard R.
United States: North Carolina, Bladen County, Elizabethtown. Cretaceous-Cretaceous Late-Campanian middle 1 rostral plate. Case *et al.*, 2012, Descr. p. 593, & fig. Fig. 2, p. 594
　　Case, Gerard R., Cook, Todd D., Wilson, Mark V. H., and Borodin, Paul D. 2012. A new species of the sclerorhynchid sawfish *Borodinopristis* from the Campanian (Upper Cretaceous) of North Carolina, USA. Historical Biology Vol. 24(6):592–597.
<u>PARATYPES:</u>
UALVP53725 one rostral plate. Case *et al.*, 2012, Descr. p. 593, & fig. Fig. 3, A, p. 595
UALVP53726 one rostral plate. Case *et al.*, 2012, Descr. p. 593, & fig. Fig. 3, B, p. 595
UALVP53727 one rostral plate. Case *et al.*, 2012, Descr. p. 593, & fig. Fig. 3, C, p. 595

Chondrichthyes Huxley, 1880
　　Elasmobranchii Bonaparte, 1838
　　　　Batomorphii Cappetta, 1980
　　　　　　Rajiformes Berg, 1940
　　　　　　　　Sclerorhynchoidei Cappetta, 1980
　　　　　　　　　　Ptychotrygonidae Kriwet, Nunn, and Klug, 2009

UALVP57039 HOLOTYPE *Ptychotrygon clementsi* **Case, Cook, Sadorf, and Shannon, 2017**
Collectors: Case, Gerard R., Sadorf, Eric M., and Shannon, Kevin R. United States: North Carolina: Pender County: Castle Hayne, Martin-Marietta Castle Hayne Quarry, 34°22′01.87″N. Lat. 77°51′ 07.07″W. Long. Cretaceous-Cretaceous Late-Maastrichtian. Peedee Formation, Island Creek Member. Complete tooth of indeterminate jaw position. Case *et al.*, 2017, Descr. pp. 74–75, & fig. Fig. 4, F, p. 73

Case, Gerard R., Cook, Todd D., Sadorf, Eric M., and Shannon, Kevin R. 2017. A late Maastrichtian selachian assemblage from the Peedee Formation of North Carolina, USA. Vertebrate Anatomy Morphology Palaeontology, Vol. 3:63–80.
PARATYPE:
UALVP57040 complete tooth of indeterminate jaw position. Case *et al.*, 2017, Descr. pp. 74–75, & fig. Fig. 4, G, p. 73

Chondrichthyes Huxley, 1880
   Elasmobranchii Bonaparte, 1838
      Eugeneodontioformes Zangerl, 1981
         Edestidae Jaekel, 1899
UALVP46579 HOLOTYPE *Paredestus bricircum* **Mutter and Neuman, 2008, p. 25**
Collector: Lindoe, Allan L., 2003.
Canada: British Columbia, Wapiti Lake, Wapiti Lake C.
Triassic-Triassic Lower-Smithian, Sulphur Mountain Fm. Dentition, tooth whorl, jaw, six parts. Mutter and Neuman, 2008, Descr. pp. 25–27, p. 33, & fig. Fig. 19, A, B, p. 27. Ginter *et al.*, 2010. Handbook of Paleoichthyol. Vol. 3D: fig. Fig. 129, p. 133. CCIS L2-245, Triassic 05, 08

Mutter, R. J., and Neuman, A. G. 2008. New eugeneodontid sharks from the Lower Triassic Sulphur Mountain Formation of Western Canada. pp. 9–41. *IN:* Cavin, L., Longbottom, A., and Richter, M. (editors). Fishes and the Break-Up of Pangea. Geological Society Special Publication No. 295. London, England, U.K. 372 pp.

Chondrichthyes Huxley, 1880
Elasmobranchii Bonaparte, 1838
Wapitiodidae
Wapitiodus aplopagus Mutter, De Blanger, and Neuman, 2007.
Canada: British Columbia, Wapiti Lake, Wapiti Lake C. Triassic-
Triassic Early-Smithian, Sulphur Mountain Fm.

Mutter, R. J., De Blanger, K., and Neuman, A. G. 2007.
Elasmobranchs from the Lower Triassic Sulphur Mountain
Formation near Wapiti Lake (BC, Canada). Zoological Journal of
the Linnean Society, Vol. 149:309–337.

PARATYPES:
UALVP17932 fin spine, Mutter *et al.*, 2007, Descr. p. 313, teaching
collection, Z-425
UALVP46527 part of body, Mutter *et al.*, 2007, fig. Fig. 12, p. 322.
CCIS L2-245, Triassic 07, 11
UALVP46528 part of body, Mutter *et al.*, 2007, fig. Fig. 8, A, p. 319
UALVP46529 anterior body preserved in ventral view with pec-
toral fins. Mutter *et al.*, 2007, fig. Fig. 10, Figs. 11, A, B, p. 321. CCIS
L2-245, Triassic 07, 11

Grade Teleostomi
Acanthodii Owen, 1846
UALVP32454 HOLOTYPE *Cassidiceps vermiculatus* Hanke,
2001, p. 389
Collected 1990.
Canada: Northwest Territories, Mackenzie Mountains, MOTH:
Man on the Hill #1
Devonian-Devonian Early-Lochkovian, Road River Formation,
fish skeleton, part and counterpart; type tecto rostral plate.
Hanke, 2001, Descr. p. 389, fig. Fig. 126, p. 392, Fig. 127, p. 394,
Fig. 128.1–6, p. 396, Fig. 129.1–11, p. 399, Fig. 130.1–3, p. 401. CCIS
L2-245, Devonian 14, 12

Hanke, Gavin F. 2001. Comparison of an Early Devonian
Acanthodian and Putative Chondrichthyan assemblage using
both isolated and articulated remains from the Mackenzie

Mountains, with a cladistic analysis of Early Gnathostomes. Ph.D. thesis, Department of Biological Sciences, University of Alberta, Edmonton, Alberta, Canada 566 pp.

PARATYPE:
UALVP45213 isolated head scale. Hanke, 2001, Descr. p. 389, fig. Fig. 129.12, p. 399

Grade Teleostomi
   Acanthodii Owen, 1846
UALVP41484 HOLOTYPE *Ornatacanthus braybrooki* Hanke, 2001, p. 237
Collector: Hanke, Gavin, 1996.
Canada: Northwest Territories, Mackenzie Mountains, MOTH: Man on the Hill #1
Devonian-Devonian Early-Lochkovian, Road River Formation. Articulated scale patch of *Lupopsyrus*-like fish, two dorsal spines, anal spine, pelvics, two pairs of intermediate spines and broken pectoral spines present. Hanke, 2001, Descr. p. 237–252, fig. Fig. 69, p. 241, Fig. 70, p. 243, Fig. 71.1–4, p. 246, Fig. 72.1–8, p. 249, Fig. 73.1–5, p. 251

   Hanke, Gavin F. 2001. Comparison of an Early Devonian Acanthodian and Putative Chondrichthyan assemblage using both isolated and articulated remains from the Mackenzie Mountains, with a cladistic analysis of Early Gnathostomes. Ph.D. thesis, Department of Biological Sciences, University of Alberta, Edmonton, Alberta, Canada 566 pp.

Grade Teleostomi
   Acanthodii Owen, 1846
      Acanthodiformes Berg, 1940
         Mesacanthidae Moy-Thomas, 1939
UALVP41860 HOLOTYPE *Promesacanthus eppleri* Hanke, 2008, p. 292
Collector: Alan Lindoe, 1996

Canada: Northwest Territories, Mackenzie Mountains, MOTH: Man on the Hill #1

Devonian-Devonian Early-Lochkovian, Road River Formation. Acanthodian resembling *Mesacanthus*, scales missing off parts of flank, and details of the head are indistinct. Hanke, 2001, Paratype for *Promesacanthus hundaae*, Hanke, 2001, Ph.D. thesis, Descr. p. 404, fig. Fig. 131.1, p. 406, Fig. 132.1, p. 408, Hanke, 2008, Descr. p. 292, p. 297, & fig. Fig. 2, A and B, p. 292, as *Promesacanthus eppleri*. CCIS L2-245, Devonian 14, 12; CCIS L2-245, Devonian 05, 03

Hanke, Gavin F. 2008. *Promesacanthus eppleri* n. gen., n. sp., a mesacanthid (Acanthodii, Acanthodiformes) from the Lower Devonian of Northern Canada. Geodiversitas Vol. 30(2): 287–302

PARATYPE:

UALVP41672 Hanke, 2001, Ph.D. thesis, Paratype for *Promesacanthus hundaae*, Hanke, 2001, Ph.D. thesis, Descr. p. 404, & fig. Fig. 136.1–4, 6, p. 418, Fig. 137.3, 137.6, p. 420, Hanke, 2008, Descr. p. 292, & fig. Fig. 6, C, F, p. 297, Fig. 7, C, D, E–G, p. 299, as *Promesacanthus eppleri*. CCIS L2-245, Devonian 14, 12

UALVP42152 scales. Not listed in material examined but cited in Hanke, 2008, fig. Fig. 4, C, D, p. 294, Fig. 5, A, B, D, E, p. 295, as *Promesacanthus eppleri*

UALVP42651 poorly preserved, headless specimen. Hanke, 2008, Descr. p. 292, as *Promesacanthus eppleri*. CCIS L2-245, Devonian 14, 12

UALVP42652 well-preserved specimen, pectoral girdle to tail, all spines visible. Hanke, 2001, Ph.D. thesis, Paratype for *Promesacanthus hundaae*, Descr. p. 404, fig. Fig. 131.2, p. 406, Fig. 132.2, p. 408, Fig. 136.5 p. 418, Fig. 137.1–2, 4, p. 420, Fig. 139.2–10, p. 426, Hanke, 2008, Descr. p. 292, p. 293, p. 297, & fig. Fig. 3, A, B, p. 293, Fig. 6, A, B, D, p. 297, as *Promesacanthus eppleri*. CCIS L2-245, Devonian 14, 12

UALVP42653 poorly preserved specimen, from pectoral girdle to anal fin remaining. Hanke, 2001, Ph.D. thesis, Paratype for

Promesacanthus hundaae, Descr. p. 404, Hanke, 2008, Descr. p. 292, p. 297, as *Promesacanthus eppleri.* CCIS L2-245, Devonian 14, 12

UALVP43027 head and pectoral girdle of a *Mesacanthus*-like acanthodian. Hanke, 2001, Ph.D. thesis, Holotype of *Promesacanthus hundaae,* Descr. p. 404, fig. Fig. 133.1, p. 410, Fig. 134.1, p. 412, Fig. 135.4, 135.6, p. 415, Hanke and Wilson, 2006, JVP Vol. 26(3):526–537, p. 529, as *Brochoadmones milesi,* Hanke, 2008, Descr. p. 292, p. 293, & fig. Fig. 4, A, B, p. 294, Fig. 5, F, p. 295, as *Promesacanthus eppleri.* CCIS L2-245, Devonian 14, 12

Grade Teleostomi
 Acanthodii Owen, 1846
  Diplacanthiformes Berg, 1940
   Tetanopsyridae Gagnier, Hanke, and Wilson, 1999, p. 83

UALVP43246 HOLOTYPE *Tetanopsyrus breviacanthias* **Hanke, Davis, and Wilson, 2001, p. 743**
Collector: Lindoe, L. Allan, 1998.
Canada: Northwest Territories, Mackenzie Mountains, MOTH: Man on the Hill #1. Devonian-Devonian Early-Lochkovian, Delorme Group. Best preserved of this species; preserved in right, lateral view. Hanke, 2001, Ph.D. thesis, Descr. p. 324, & fig. Fig. 101, p. 326, Fig. 102, p. 329, Fig. 103.1, p. 331, Fig. 103.5, p. 331, Fig. 104.3, p. 323, Fig. 105.1, 105.5, 105.7, p. 335, Fig. 106.1–2, p. 338, Hanke *et al.*, 2001, fig. Fig. 4, A, B, p. 744; Fig. 5, B, p. 745; Fig. 6, A, p. 746, Fig. 7, A, B, C, p. 747; ratio of pectoral spine length to pelvic spine length Fig. 9, p. 749. CCIS L2-245, Devonian 14, 11

 Hanke, Gavin F., Davis, Samuel P., and Wilson, Mark V. H. 2001. New Species of the acanthodian genus *Tetanopsyrus* from Northern Canada, and comments on related taxa. Journal of Vertebrate Paleontology Vol. 21(4):740–753.
PARATYPES:
UALVP39062 on slab with UALVP32672; Gagnier *et al.* 1998. Ichthy. Issues Spec. Publication 4, pp. 12–14. Gagnier *et al.* 1999,

Acta Geologica Polonica Vol. 49(2):81–96, Descr. p. 84, fig. Fig. 8, b, p. 91, Hanke, 2001, Ph.D. thesis, Descr. p. 324, fig. Fig. 104, 1, 104.4, p. 333, Fig. 105.2, 105.6, 105.8, p. 335, Hanke *et al.*, 2001, fig. Fig. 6, C, E, p. 746; Fig. 5, B, p. 745; Fig. 7, D, p. 747; Fig. 9, p. 749. CCIS L2-245, Devonian 14, 04

UALVP42512 mostly full body, preserved in right, lateral view; missing part of rostrum; Gagnier *et al.* 1998. Ichthy. Issues Spec. Publication 4, pp. 12–14; Gagnier *et al.* 1999, Acta Geologica Polonica Vol. 49(2):81–96, as Paratype for *Tetanopysrus lindoei*, Descr. p. 84, fig. Fig. 6, A, B, p. 88. Table 1, p. 85, Hanke, 2001, Ph.D. thesis, Descr. p. 324, fig. Fig. 103.2–4, 103.6, p. 331, Fig. 104.2, p. 333, Fig. 105.3, p 335, Fig. 106.3–4, p. 338, Hanke *et al.*, 2001, fig. Fig. 4, E, p. 744; Fig. 5, A, p. 745; Fig. 6, B, D, p. 746; Fig. 7, E, p. 747; Fig. 8, A, p. 747; Fig. 9, p. 749. CCIS L2-245, Devonian 14, 11

UALVP43030 juvenile specimen. Hanke, 2001, Ph.D. thesis, Descr. p. 324, Hanke *et al.*, 2001, Descr. pp. 740–753. fig. Fig. 5, C, p. 745, Descr. p. 746 although UALVP43030 not listed in referred material on p. 743.

UALVP43089 juvenile preserved on its left side; on the same block as UALVP43090. Hanke, 2001, Ph.D. thesis, Descr. p. 324, Hanke *et al.*, 2001, fig. Fig. 5, C, p. 745; Fig. 7, F, p. 747; Fig. 9, p. 749. CCIS L2-245, Devonian 14, 11

UALVP44030 juvenile preserved on the right side. Hanke *et al.*, 2001, Descr. pp. 743–744, p. 746, p. 748, & fig. Fig. 5, C, p. 745; ratio of pectoral spine length to pelvic spine length, Fig. 9, p. 749

UALVP45153 whole skeleton; Hanke, 2001, Ph.D. thesis, Descr. p. 324, Hanke *et al.*, 2001, Descr. p. 744, & fig. Fig. 4, p. 744; ratio of pectoral spine length to pelvic spine length, Fig. 9, p. 749. CCIS L2-245, Devonian 14, 10

Grade Teleostomi
    Acanthodii Owen, 1846
        Diplacanthiformes Berg, 1940
            Tetanopsyridae Gagnier, Hanke, and Wilson, 1999, p. 83

UALVP39078 HOLOTYPE *Tetanopsyrus lindoei* **Gagnier, Hanke, and Wilson, 1999, p. 83**
Collector: Mark V. H. Wilson, 1990
Canada: Northwest Territories, Mackenzie Mountains, MOTH: Man on the Hill #1
Devonian-Devonian Early-Lochkovian, Road River Formation. A specimen preserved in right lateral view; on slab with UALVP38678-UALVP38682; interesting spine (segmented lepidotrochia-like structure); Gagnier *et al.*, 1998, Ichthy. Issues Spec. Publication 4, pp. 12–14; Gagnier *et al.*, 1999, Vol. 49(2):81–96, Descr. p. 84, Table 1, p. 85, & fig. Fig. 4, a, b, p. 86, Fig. 5, a, b, p. 87, Hanke, 2001, Ph.D. thesis, Descr. p. 312, & fig. Fig. 99.1–2, 99.4–5, p. 321, Fig. 100.2–3, p. 323, Hanke *et al.*, 2001, JVP 21(4):740–753 Fig. 3, E, p. 743; Fig. 3, E, p. 743. CCIS L2-245, Devonian 14, 04. CCIS L2-245, Devonian 14, 05

> **Gagnier, P.-Y., Hanke, Gavin F., and Wilson, Mark V. H. 1999.** *Tetanopsyrus lindoei*, gen et sp. nov., an Early Devonain acanthodian from the Northwest Territories. Acta Geologica Polonica Vol. 49:81–96.

PARATYPES:
UALVP38682 complete fish lacking most of the body scales (on slab with UALVP39078 & UALVP38678-UALVP38682); Gagnier *et al.*, 1998, Ichthy. Issues Spec. Publication 4, pp. 12–14, *gen. nov. sp. nov.*, type buccal plate, Gagnier *et al.*, 1999, Descr. p. 84, Table 1, p. 85, p. 92, & Fig. 8,A, p. 91, Hanke, 2001, Ph.D. thesis, Descr. p. 312, Hanke *et al.*, 2001, Descr. pp.740–753, & fig. Fig. 9. p. 749. CCIS L2-245, Devonian 14, 05
UALVP39062 an articulated specimen lacking the head and the tip of the tail on slab with UALVP32672; Gagnier *et al.*, 1998, Ichthy. Issues Spec. Publication 4, pp. 12–14. Gagnier *et al.*, 1999, Descr. p. 84, Fig. 8b, p. 91, Hanke, 2001, Ph.D. thesis, Descr. p. 324, fig. Fig. 104, 1, 104.4, p. 333, Fig. 105.2, 105.6, 105.8, p. 335, Hanke *et al.*, 2001, fig. Fig. 6, C, E, p. 746; Fig. 5, B, p. 745; Fig. 7, D, p. 747; Fig. 9, p. 749, Paratype for *Tetanopsyrus lindoei* Gagnier

et al., 1999, now identified as *Tetanopsyrus breviacanthias*. CCIS L2-245, Devonian 14, 04

UALVP39084 full body with head partially eroded. Gagnier *et al.* 1998, Ichthy. Issues Spec. Publication 4, pp. 12–14, Gagnier *et al.* 1999, Descr. p. 84, Table 1, p. 85, Fig. 7, A, (misprinted as UALVP42512 in that figure), p. 90, Hanke *et al.*, 2001, Descr, p. 743, Paratype for *Tetanopsyrus lindoei*, Gagnier *et al.*, 1999, but reidentified in Hanke *et al.*, 2001, as *Tetanpysyrus* sp.

UALVP42512 mostly full body, preserved in right, lateral view; missing part of rostrum; Gagnier *et al.*, 1998, Ichthy. Issues Spec. Publication 4, pp. 12–14, Gagnier *et al.* 1999, as Paratype for *Tetanopysrus lindoei*, Descr. p. 84, & fig. Fig. 6, A, B, p. 88. Table 1, p. 85, Hanke, 2001, Ph.D. thesis, Descr. p. 324, fig. Fig. 103.2–4, 103.6, p. 331, Fig. 104.2, p. 333, Fig. 105.3, p 335, Fig. 106.3–4, p. 338, Hanke *et al.*, 2001, fig. Fig. 4, E, p. 744; Fig. 5, A, p. 745; Fig. 6, B, D, p. 746; Fig. 7, p. 747; Fig. 8, A, p. 747; Fig. 9, p. 749, now identified as *Tetanopsyrus breviacanthias*. CCIS L2-245, Devonian 14, 11

UALVP42648 headless specimen, spines identical to gen. et sp. nov. (see Gagnier and Wilson 1995), pelvic spine and scales taken for sectioning. Gagnier *et al.* 1999, Descr. p. 84, Hanke *et al.*, 2001, p. 743. CCIS L2-245, Devonian 14, 02

Grade Teleostomi
 Acanthodii Owen, 1846
  Ischnacanthiformes Berg, 1940
   Ischnacanthidae Woodward, 1891
UALVP42666 HOLOTYPE *Erymnacanthus clivus* Blais, **Hermus, and Wilson, 2015**
Collector: Lindoe, L. Allan. 1996.
Canada: Northwest Territories, Mackenzie Mountains, MOTH: Man on the Hill #1
Devonian-Devonian Early-Lochkovian, Road River Formation. Set of dentigerous jaw bones: nearly complete right upper dentigerous jaw bones & palatoquadrate cartilage in lateral view,

complete left & right lower dentigerous jaw bones and Meckel's cartilages preserved in lingual view. Published as *Ischnacanthus* in Hanke *et al.*, 2001, Can. J. Earth Sci., Vol. 38 p. 1525, Blais *et al.*, 2015, fig. Fig. 4A, B, p. 7. Teaching Collection L4-b. Devonian 15, 19

Blais, Stephanie, Hermus, C., and Wilson, Mark V. H. 2015. Four new Early Devonian ischnacanthid acanthodians from the Mackenzie Mountains, Northwest Territories, Canada: an early experiment in dental diversity. Journal of Vertebrate Paleontology, Vol. 35(1): 1–13.

PARATYPES:

UALVP42198 upper left dentigerous jaw bones & palatoquadrate cartilages in lingual view. Blais *et al.*, 2015, Descr. p. 6

UALVP45077 upper left dentigerous jaw bone & palatoquadrate cartilage in lingual view. Blais *et al.*, 2015, Descr. p. 6

UALVP45097 partly disarticulated specimen showing neural arches, jaws/teeth, fin spines, body scales. Blais *et al.*, 2015, Descr. p. 6

UALVP47234 Partial upper right dentigerous jaw bone & palatoquadrate cartilage in lingual view. Blais *et al.*, 2015, Descr. p. 6

Grade Teleostomi
    Acanthodii Owen, 1846
        Ischnacanthiformes Berg, 1940
            Ischnacanthidae Woodward, 1891
UALVP45648 HOLOTYPE *Euryacanthus rugosus* **Blais, Hermus, and Wilson, 2015**
Collector: Hanke, Gavin, 2002.
Canada: Northwest Territories, Mackenzie Mountains, MOTH: Man on the Hill #1, Devonian-Devonian Early-Lochkovian, Delorme Group, Lithology: limestone. Large right upper and lower jaws. Blais *et al.*, 2015, fig. Fig. 3, A–C, p. 5 Devonian 15, 19

Blais, Stephanie, Hermus, C., and Wilson, Mark V. H. 2015. Four new Early Devonian ischnacanthid acanthodians from the Mackenzie Mountains, Northwest Territories, Canada: an early experiment in dental diversity. Journal of Vertebrate Paleontology, Vol. 35(1): 1–13

PARATYPES:

UALVP41650 anterior half of a lower right jaw bone & Meckel's cartilage in lingual view Blais *et al.*, 2015 Descr. p. 4

UALVP42023 right upper dentigerous jaw bone & palatoquadrate cartilage. Blais *et al.*, 2015, Descr. p. 4, Teaching Collection, L4-a

UALVP42025 left upper dentigerous jaw bone & palatoquadrate cartilage in lingual view. Blais *et al.*, 2015, Descr. p. 4, Teaching Collection, L4-a

UALVP45040 isolated jaw element; very robust with more than eight teeth, right upper dentigerous bone & palatoquadrate cartilage in lingual view. Blais *et al.*, 2015, fig. Fig. 3, D, p. 5. Devonian 15, 19

UALVP45076 (=UALVP41663) isolated jaw element, partial lower left jawbone & Meckel's cartilage in lingual view. This specimen was previously catalogued as UALVP41663. The specimen number was removed during preparation and given the new number of UALVP45076. Blais *et al.*, 2015, Descr. p. 4. Teaching Collection, L4-a. Devonian 15, 19

Grade Teleostomi
    Acanthodii Owen, 1846
        Ischnacanthiformes Berg, 1940
        Ischnacanthidae Woodward, 1891

UALVP56502 HOLOTYPE *Euryacanthus serratus* **Blais and Wilson, (submitted)**

Collector: Lindoe, L. Allan, 2013.

Canada: Northwest Territories, B-MOTH (?=GSC 69063), Silurian, Late Wenlockian (Homerian) or early Ludlovian (Gorstian), Road River Formation. Lithology: Argillaceous limestone/calcarous fossiliferous siltstone. Right upper palatoquadrate cartilage and dentigerous jaw bone in lingual view

    Blais, Stephanie, and Wilson, Mark V. H. (submitted). Two new Silurian ischnacanthiforms from the Northwest Territories of Canada and evidence for tooth-on-tooth wear in early jawed vertebrates. Canadian Journal of Earth Sciences.

Grade Teleostomi
    Acanthodii Owen, 1846
        Ischnacanthiformes Berg, 1940
            Ischnacanthidae Woodward, 1891
UALVP56501 HOLOTYPE *Oroichthys theobromodon* **Blais and Wilson (submitted)**
Collector: Blais, Stephanie 2013.
Canada: Northwest Territories, B-MOTH (?=GSC 69063), Silurian, Late Wenlockian (Homerian) or early Ludlovian (Gorstian), Road River Formation. Lithology: Argillaceous limestone/calcarous fossiliferous siltstone. Isolated dentigerous jaw bone fragment
    Blais, Stephanie, and Wilson, Mark V. H. (submitted). Two new Silurian ischnacanthiforms from the Northwest Territories of Canada and evidence for tooth-on-tooth wear in early jawed vertebrates. Canadian Journal of Earth Sciences.

Grade Teleostomi
    Acanthodii Owen, 1846
        Ischnacanthiformes Berg, 1940
            Ischnacanthidae Woodward, 1891
UALVP45078 HOLOTYPE *Tricuspicanthus gannitus* **Blais, Hermus, and Wilson, 2015**
Canada: Northwest Territories, Mackenzie Mountains, MOTH: Man on the Hill #1
Devonian-Devonian Early-Lochkovian, Road River Formation. Articulated upper & lower left dentigerous jaw bones & cartilages preserved in lingual view. Blais *et al.*, 2015, fig. Fig. 5, A, p. 8. CCIS L2-245, Devonian 15, 19
    Blais, Stephanie, Hermus, C., and Wilson, Mark V. H. 2015. Four new Early Devonian ischnacanthid acanthodians from the Mackenzie Mountains, Northwest Territories, Canada: an early

experiment in dental diversity. Journal of Vertebrate Paleontology, Vol. 35(1): 1–13

PARATYPES:

UALVP23294 lower left dentigerous jaw bone & Meckel's cartilage preserved in lingual view. Blais *et al.*, 2015, Descr. p. 7

UALVP32443 left upper dentigerous jaw bone & palatoquadrate cartilage in lingual view. Blais *et al.*, 2015, Descr. p. 7

UALVP41527 left upper (or, right lower) dentigerous jaw bone in lingual view, without associated cartilage. Blais *et al.*, 2015, Descr. p. 7

UALVP41663 small right & left upper dentigerous jaw bones & palatoquadrate cartilages in lingual view. Blais *et al.*, 2015, Descr. p. 7. Listed as paratype of Agnatha *Waengsjoeaspis platycornis*. Scott and Wilson, 2012, p. 1240. Also, see Acanthodii *Euryacanthus rugosus*. Blais *et al.*, 2015, under UALVP45076 (=UALVP41663). CCIS L2-245, Devonian 15, 19

UALVP42015 right upper jaw bone & partial palatoquadrate cartilage in lingual view. Blais *et al.*, 2015, Descr. p. 7

UALVP42062 left lower dentigerous jaw bone with Meckel's cartilage in lingual view. Blais *et al.*, 2015, Descr. p. 7. CCIS L2-245, Devonian 14, 17

UALVP42143 right lower dentigerous jaw bone with Meckel's cartilage in lingual view. Blais *et al.*, 2015, Descr. p. 7

UALVP42199 complete left & right upper dentigerous jaw bones & palatoquadrate cartilages preserved in lingural view. Blais *et al.*, 2015, Descr. p. 7. CCIS L2-245, Devonian 15, 19

UALVP42658 left upper dentigerous jaw bone & palatoquadrate cartilage in lingual view. Blais *et al.*, 2015, Descr. p. 7. CCIS L2-245, Devonian 15, 19

UALVP42659 anterior portion of mid-sized fish with right lower dentigerous bone & Meckel's cartilage visible in lingual view. Descr. as *Ischnacanthus*. Hanke *et al.*, 2001, Can. J. Earth Sci. Vol. 38 p. 1525, Blais *et al.*, 2011, p. 1191, Scale type indet., Blais *et al.*, 2015, Descr. p. 7. CCIS L2-245, Devonian 15, 19

UALVP42660 head of mid-sized fish preserved in right lateral view. Blais *et al.*, 2015, figured Fig. 6A–C, Descr. as

Ischnacanthus in Hanke *et al.,* 2001, Can. J. Earth Sci., Vol. 38: p. 1525, Blais *et al.,* 2011, Descr. p. 1191, Type A scale; Blais *et al.,* 2015, Descr. p. 7, fig. Fig. 6 A, B, C, p. 9. CCIS L2-245, Devonian 15, 19

UALVP45074 lower left dentigerous jaw bones & Meckel's cartilages in lingual view. Blais *et al.,* 2015, Descr. p. 7 Teaching Collection, L4-a

UALVP45075 left upper dentigerous jaw bones and palatoquadrate cartilages in lingual view. Blais *et al.,* 2015, Descr. p. 7. CCIS L2-245, Devonian 14, 18

UALVP45649 Right upper & lower jaw bones & associated cartilages in lingual view. Blais *et al.,* 2015, Descr. p. 7

UALVP45650 partial right upper jaw; there is another jaw on the block that is believed to be associated with the same specimen. Blais *et al.,* 2015, Descr. p. 7

Grade Teleostomi
    Acanthodii Owen, 1846
        Ischnacanthiformes Berg, 1940
        Ischnacanthidae Woodward, 1891

UALVP45072 HOLOTYPE *Tricuspicanthus pisciculus* **Blais, Hermus, and Wilson, 2015**

Canada: Northwest Territories, Mackenzie Mountains, MOTH: Man on the Hill #1

Devonian-Devonian Early, Lithology: shale. Small left upper dentigerous jaw bone & palatoquadrate cartilage preserved in lingual view. Blais *et al.,* 2015, fig. Fig.7, p. 10. Devonian 15, 19

    Blais, Stephanie, Hermus, C., and Wilson, Mark V. H. 2015. Four new Early Devonian ischnacanthid acanthodians from the Mackenzie Mountains, Northwest Territories, Canada: an early experiment in dental diversity. Journal of Vertebrate Paleontology Vol. 35(1): 1–13

PARATYPE:

UALVP45620 partial upper jaw on block with UALVP45037; small left upper dentigerous jaw bone & palatoquadrate cartilage preserved in lingual view. Blais *et al.*, 2015, Descr. p. 10. Devonian 15, 19

Grade Teleostomi
   Acanthodii Owen, 1846
      Ischnacanthiformes Berg, 1940
         Ischnacanthidae Woodward, 1891
UALVP43081 HOLOTYPE *Xylacanthus kenstewarti* **Hanke, Wilson, and Lindoe, 2001**
Collector: Gavin Hanke, 1998
Canada: Northwest Territories, southern Mackenzie Mountains, B-MOTH fish layer (Gabrielse *et al.* 1973=GSC 69063). Silurian-Silurian Early-Wenlockian Late-Homerian, Road River Formation. Lithology: calcareous fossiliferous silstone. Right lower jaw bone and partial Meckel's cartillage, in lingual view. Hanke *et al.*, 2001, fig. Fig. 3, C, p. 1521, Fig. 4, C, p. 1522. CCIS L2-245 Silurian 14, 02
   Hanke, Gavin F., Wilson, Mark V. H., and Lindoe, L. Allan. 2001. New species of Silurian acanthodians from the Mackenzie Mountains, Canada. Canadian Journal of Earth Sciences Vol. 38(11):1517–1529.
<u>**PARATYPES:**</u>
UALVP43079 left lower jaw in lingual view. Hanke *et al.*, 2001, fig. Fig. 3A, p. 1521, Fig. 4, A, G, p. 1522. CCIS L2-245 Silurian 14, 02
UALVP43080 isolated jaw bones; left, lower, lingual view. Hanke *et al.*, 2001, fig. Fig. 4, B, p. 1522. CCIS L2-245 Silurian 14, 02

Grade Teleostomi
   Acanthodii Owen, 1846
      Ischnacanthiformes Berg, 1940
         Ischnacanthidae Woodward, 1891
UALVP43242 HOLOTYPE *Granulacanthus joenelsoni* **Hanke, Wilson, and Lindoe, 2001, p. 1521**
Collector: Lindoe, L. Allan, 1998.
Canada: Northwest Territories, southern Mackenzie Mountains, B-MOTH fish layer (Gabrielse *et al.* 1973=GSC 69063).

Silurian-Silurian Early-Wenlockian Late-Homerian, Road River Formation. Lithology: calcareous siltstone. Fin spine. Hanke *et al.*, 2001, fig. Fig. 4E, 4F, p. 1522; Fig. 5B, p. 1524. CCIS L2-245, Silurian 10, 01

Hanke, Gavin F., Wilson, Mark V. H., and Lindoe, L. Allan. 2001. New species of Silurian acanthodians from the Mackenzie Mountains, Canada. Canadian Journal of Earth Sciences Vol. 38(11):1517–1529.

PARATYPES:

UALVP43241 fin spine. Hanke *et al.*, 2001, fig. Fig. 4, D, p. 1522; Fig. 5, A, p. 1524. CCIS L2-245, Silurian 10, 1
UALVP43243 fin spine. Hanke *et al.*, 2001, fig. Fig. 5, C, p. 1524
UALVP43244 fin spine. Hanke *et al.*, 2001, fig. Fig. 5, D, p. 1524

Grade Teleostomi
    Acanthodii Owen, 1846
UALVP43240 HOLOTYPE *Paucicanthus vanelsti* Hanke, 2001, p. 361
Collector: Lindoe, L. Allan, 1998
Canada: Northwest Territories, Mackenzie Mountains, MOTH: Man on the Hill #1. Devonian-Devonian Early-Lochkovian, Delorme Group. Nearly complete specimen; preserved in right, lateral view. Hanke, 2001, Descr. p. 361, & fig. Fig. 116.1, p. 364. Fig. 117.1, p. 366, Fig. 118.1, p. 368, Fig. 119.3–4, p. 371, Fig. 120.1, 120.3, 120.6, p. 373. CCIS L2-245, Devonian 14, 14

Hanke, Gavin F. 2001. Comparison of an Early Devonian Acanthodian and Putative Chondrichthyan assemblage using both isolated and articulated remains from the Mackenzie Mountains, with a cladistic analysis of Early Gnathostomes. Ph.D. thesis Department of Biological Sciences, University of Alberta, Edmonton, Alberta, Canada 566 pp.

PARATYPES:

UALVP41932 Hanke, 2001, Descr. p. 361. CCIS L2-245, Devonian 14, 13

UALVP42160 Hanke, 2001, Descr. p. 361. CCIS L2-245, Devonian 14, 13

UALVP43410 full body preserved in left, lateral view. Hanke, 2001, Descr. p. 361, fig. Fig. 116.2, p. 364, Fig. 117.2, p. 366, Fig. 118.2, p. 368, Fig. 119.1–2, Fig. 119.5, p. 371, Fig. 120.2, 120.4–5, p. 373. CCIS L2-245, Devonian 14, 14

UALVP44045 Hanke, 2001, Descr. p. 361. CCIS L2-245, Devonian 14, 12

Grade Teleostomi
    Acanthodii Owen, 1846
        Lupopsyriformes Hanke, 2001
        Lupopsyridae Hanke, 2001

UALVP43009 HOLOTYPE *Lupopsyroides macracanthus* **Hanke and Wilson, 2004**

Collector: Lindoe, L. Allan, 1998

Canada: Northwest Territories, Mackenzie Mountains, MOTH: Man on the Hill #1. Devonian-Devonian Early-Lochkovian, Road River Formation. Acanthodian resembles *Lupopsyrus*, though prepelvic spines like *Kathemacanthus*. Hanke, 2001, Ph.D. thesis, Descr. pp. 46–58, p. 47, Fig. 9, p. 49, Fig. 10, p. 51, Fig. 11(1–6), p. 54. Hanke and Wilson, 2004, Descr. pp. 189–215. CCIS L2-245, Devonian 14, 07

    Hanke, Gavin F., and Wilson, Mark V. H. 2004. New teleostome fishes and acanthodian systematics. pp. 189–215. *IN:* Arratia, G., Wilson, Mark V. H., and Cloutier, R. (editors). Recent advances in the origin and early radiation of vertebrates. Dr. Friedrich Pfeil, München, Germany. 703 pp.

PARATYPES:

UALVP42532 similar in appearance to *Lupopsyroides*, but may be a different species. Hanke, 2001, Ph.D. thesis, p. 52. Hanke and Wilson, 2004. CCIS L2-245, Devonian 14, 07

UALVP45295 isolated scale. Hanke, 2001, Ph.D. thesis, p. 52, & fig. Fig. 144.10, p. 447. Hanke and Wilson, 2004

UALVP45296 isolated scale. Hanke, 2001, Ph.D. thesis, p. 52, & fig. Fig. 144.11, p. 447. Hanke and Wilson, 2004

Osteichthyes Huxley, 1880
  Sarcopterygii Romer, 1955
    Dipnoi Müller, 1844
      Dipteroidei Vorobyeva and Obruchev, 1964
        Dipteridae Owen, 1846
UALVP477 HOLOTYPE *Sunwapta grandiceps* Thomson, 1967, p. 2
Collector: Bleuler, M., 1937
Canada: Alberta, Mount Athabasca, Sunwapta Pass, Sullivan Creek. Devonian-Devonian Late. Symphysis & anterior rami of lower jaws and tooth plates. Thomson, 1967, Descr. p. 2, pp. 4–5, & Fig. 1, p. 3. Earth Sciences Museum ESB B-01; cast in Teaching collection Z-425 Unit 13, Drawer 5
    Thomson, Keith Stewart. 1967. A new genus and species of marine Dipnoan fish, from the Upper Devonian of Canada. Postilla No. 106:1–6

Osteichthyes Huxley, 1880
  Sarcopterygii Romer, 1955
    Actinistia Cope, 1871
      Coelacanthoidei Berg, 1937
        Whiteiidae Schultze, 1993
RBCM.EH 1986.001.0022 HOLOTYPE at Royal British Columbia Museum, Victoria, British Columbia. *Whiteia lepta* **Wendruff, 2011, p. 88**
Canada: British Columbia, Wapiti Lake, commercial quarry in Cirque C. 54°30′N. Lat., 120°43′W. Long. Triassic-Triassic Lower-Smithian, Sulphur Mountain Formation. A complete and articulated specimen from the skull to the tip of the supplementary lobe, skull poorly preserved
    Wendruff, Andrew James 2011. Lower Triassic Coelacanths of the Sulphur Mountain Formation (Wapiti Lake) in British

Columbia, Canada. University of Alberta, Department of Biological Sciences. M.Sc. thesis, Edmonton, Alberta 304 pp.
PARATYPES:
UALVP43382 anterior half of skeleton from the skull to the pelvic girdle. Wendruff, 2011, Descr. p. 82, p. 85, p. 95, p. 96, p. 97, p. 98, p. 104, & fig. Fig. 3.3, A, B, p. 147. CCIS L2-245 Triassic 04, 16
UALVP43602 complete skull in dorsolateral view with partial pectoral fin, cheek bones mostly missing, nearly complete skull roof in dorsal view, length 26 cm. Wendruff, 2011, Descr. p. 85, p. 91, p. 92, p. 93, p. 94, p. 95, p. 96, p. 97, p. 98, & fig. Fig. 3.2, F, p. 145. CCIS L2-245 Triassic 07, 03
UALVP43719 partial caudal fin, 18 cm deep. Wendruff, 2011, Descr. p. 82. CCIS L2-245 Triassic 06, 15

Osteichthyes Huxley, 1880
    Sarcopterygii Romer, 1955
        Rhipidistia Cope, 1889
UALVP24228 HOLOTYPE *Wapitia robusta* **Wendruff. 2011, p. 191**
Collector: Lindoe, L. Allan, and Neuman, Andrew G.
Canada: British Columbia, Wapiti Lake. 54°32′N. Lat., 120°45′24″W. Long. Triassic-Triassic Early-Smithian, Sulphur Mountain Formation. Complete specimen. Wendruff, 2011, Descr. p. 190, p. 192, p. 193, p. 194, p. 195, p. 196, p. 197, p. 198, p. 199, p. 200, p. 201, & fig. Fig. 4.8, p. 239, Fig. 4.9, A, B, C, p. 241. Earth Sciences Museum, ESB B-01
    Wendruff, Andrew James 2011. Lower Triassic Coelacanths of the Sulphur Mountain Formation (Wapiti Lake) in British Columbia, Canada. University of Alberta, Department of Biological Sciences. M.Sc. thesis, Edmonton, Alberta, 304 pp.
PARATYPES:
UALVP43604 isolated scale. Wendruff, 2011, Descr. p. 191, & fig. Fig. 4.11, A, p. 245
UALVP43605 isolated complete scale. Wendruff, 2011, Descr. p. 191, & fig. Fig. 4.11, B, p. 245

Osteichthyes Huxley, 1880
   Sarcopterygii Romer, 1955
      Rhipidistia Cope, 1889
         Actinistia Cope, 1871
TMP 1995.118.23 HOLOTYPE at Royal Tyrrell Museum of Paleontology. *Everticauda pavoidea* **Wendruff, 2011, p. 172**
Canada: British Columbia, Wapiti Lake, commercial quarry in Cirque C. 54°30′N. Lat., 120°43′W. Long. Triassic-Triassic Lower-Smithian, Sulphur Mountain Formation. An articulated specimen, missing the skull and tips of the caudal fin but otherwise nearly complete

   Wendruff, Andrew James 2011. Lower Triassic Coelacanths of the Sulphur Mountain Formation (Wapiti Lake) in British Columbia, Canada. University of Alberta, Department of Biological Sciences. M.Sc. thesis, Edmonton, Alberta 304 pp.
<u>**PARATYPES:**</u>
UALVP19237 articulated partial fish from pelvic girdle to anterior position of caudal fin. Wendruff, 2011, Descr. p. 172, p. 182, p. 184, p. 185, p. 188, p. 189, & fig. Fig. 4.5, p. 233. Teaching Collection, L6-b
UALVP43698 nearly complete caudal fin with supplementary lobe. Wendruff, 2011, Descr. p. 172, p. 187, p. 188, & fig. Fig. 4.4, p. 231. CCIS L2-245, Triassic 06, 10
UALVP46608 complete juvenile specimen, caudal fin missing, skull slightly disarticulated. Wendruff, 2011, Descr. p. 175, p. 176, p. 177, p. 178, p. 179, p. 180, p. 181, p. 182, p. 183, p. 184, p. 185, p. 186, p. 214, p. 215, & fig. Fig. 4.2, C, p. 227, Fig. 4.3, A, B, p. 229

Osteichthyes Huxley, 1880
   Sarcopterygii Romer, 1955
      Actinistia Cope, 1871
         Coelacanthoidei Berg, 1937
            Laugiidae Berg, 1940
UALVP43606 HOLOTYPE *Belemnocerca prolata* **Wendruff and Wilson, 2013. P. 905**

Collector: Wilson, Mark V. H., 2003.
Canada: British Columbia, Wapiti Lake, Cirque C. 54°32'N. Lat., 120°45'24"W. long. Triassic-Triassic Lower-Smithian, Sulphur Mountain Formation. Posterior body, three parts. Isolated scales. Wendruff, 2011, Descr. p. 253, & fig. Fig. 5.1, p. 268, Fig. 5.2, p. 270, Fig. 6.1, D, p. 292, Fig. 7.1, C, p. 304. Wendruff and Wilson, 2013, Descr. 905–908, Fig. 2, A, B, p. 906, Fig. 3, A, B, p. 907. CCIS L2-245, Triassic 06, 10

Wendruff, Andrew J., and Wilson, Mark V. H. 2013. New Early Triassic coelacanth in the family Laugiidae (Sarcopterygii: Actinistia) from the Sulphur Mountain Formation near Wapiti Lake, British Columbia, Canada. Canadian Journal of Earth Science, Vol. 50:904–910

Osteichthyes Huxley, 1880
    Sarcopterygii Romer, 1955
        Rhipidistia Cope, 1887
            Tetrapodomorpha Ahlberg, 1991
            Megalichthyidae Hay, 1902
UALVP45582 CAST OF HOLOTYPE, original specimen in Queensland Museum, Queensland, Australia. *Cladarosymblema narrienense* **Fox, Campbell, Barwick, and Long, 1995**
Collector: Richard C. Fox
Australia: Queensland. Carboniferous-Carboniferous Early. Two rubber casts of skull, one original shape, one with left gular bones removed. Biological Sciences Building Z-425, Unit 14, Drawer 3

Fox, R.C., Campbell, K. S. W., Barwick, R. E., and Long, J. A. 1995. A new osteolepiform fish from the Lower Carboniferous Raymond Basin, Queensland. Memoirs of the Queensland Museum, Vol. 38(1): 99–221

Osteichthyes Huxley, 1880
    Sarcopterygii Romer, 1955
        Rhipidistia Cope, 1887

Tetrapodomorpha Ahlberg, 1991
Elpistostegalia Camp and Allison, 1961
Panderichthyidae Vorobyeva, 1968

UALVP52138 CAST OF HOLOTYPE, NUFV108, Nunavut Fossil Vertebrate Collection *Tiktaalik roseae* Daeschler, Shubin, and Jenkins, Jr., 2006, p. 759

Canada: Nunavut, Ellesmere Island, southern Ellesmere Island; near the eastern arm of Bird Fiord, "Fram Valley" Site NV2K17. N. Lat. 77°09.898′, W. Long. 86°16.151′. Devonian-Devonian Late; Early to Middle Frasnian, Okse Bay Group, middle part of the Fram Formation. Skull & partial body Cast of NUFV108, Daeschler *et al.*, 2006, Descr. pp. 759–761, & fig. Fig. 2, A–C, p. 758. Fig. 3, p. 759, Fig. 6, A, B, p. 760. Z-425, teaching collection

Daeschler, E. B., Shubin, N. H., and Jenkins, Jr., F. A. 2006. A Devonian tetrapod-like fish and the evolution of the tetrapod body plan. Nature 440(4639):757–763.

Osteichthyes Huxley, 1880
Holostei Müller, 1845
Halecomorphi Cope, 1872
Amiidae Bonaparte, 1838

UALVP131 HOLOTYPE *Stylomyleodon lacus* Russell, 1928 Reidentified as *Cyclurus lacus* (Russell, 1928) in Gaudant, 1992

Collector: Russell, Loris S. September 1924

Canada: Alberta, Red Deer River, Red Deer River #13. Tertiary-Paleocene, Paskapoo Formation. Partial left splenial with grinding teeth. Russell, 1928, Descr. p. 103 & fig. Fig. 1, A, B, p. 104; Reidentified as *Cyclurus lacus* (Russell, 1928) by Gaudant, 1992. Grande and Bemis, 1998, SVP Memoir 4:1–690. As *Cyclurus lacus*, Descr p. 315, & fig. Fig. 206, A, p. 315. CCIS L2-245, Paleocene 02, 13

Russell, Loris S. 1928. A new fossil fish from the Paskapoo beds of Alberta. American Journal of Science Fifth Series, Vol. 15(86):103–107.

Gaudant, J. 1992. *"Kindleia" fragosa* Jordan and *"Stylomyleodon" lacus* Russell: two amiid fishes from the Late Cretaceous and Paleocene of Alberta, Canada. Canadian Journal of Earth Science, Vol. 29:158–173.

PARATYPES:

UALVP132 grinding teeth. Russell, 1928, Descr. p. 103 & fig. Fig. 1, C, p. 104; Reidentified as *Cyclurus lacus* (Russell, 1928) by Gaudant, 1992. Grande and Bemis, 1998, SVP Memoir 4:1–690. As *Cyclurus lacus* Descr p. 315, & fig. Fig. 206, B, p. 315. CCIS L2-245, Paleocene 02, 13

UALVP133 dentaries, scales, sculptured plates, spines, vertebrae, miscellaneous bones. Russell, 1928, Descr. p. 103, p. 106, & fig. Fig 3, 104, Fig. 4, p. 104; Reidentified as *Cyclurus lacus* (Russell, 1928) by Gaudant, 1992; Grande and Bemis, 1998, SVP Memoir 4:1–690. As *Cyclurus lacus* Descr p. 315, & fig. Fig. 206, C, D, E, F, p. 315. CCIS L2-245, Paleocene 02, 13

Osteichthyes Huxley, 1880
    Holostei Müller, 1845
        Halecomorphi Cope, 1872
            Amiidae Bonaparte, 1838

UALVP14758 HOLOTYPE *Amia hesperia* **Wilson, 1982**
Collectors: Lindoe, L. Allan, and Wilson, Mark V. H., 1977
Canada: British Columbia, Princeton #4, Tertiary-Eocene-Eocene middle, Allenby Formation. Near complete, partially articulated, part & counterpart, Wilson, 1982, Descr. pp. 413–424. Grande and Bemis, 1998, Descr. pp. 172–185; Table 30, p. 172, Tables 31, 32, p. 173, Tables 33, 34, p. 177; Tables 35, 36, p. 182, Tables 37, 38, p. 183, Table 39, p. 184. & fig. Figs. 108 A, B, p. 174, Figs. 109 A, B, p. 175, Fig. 111, p. 178, Fig. 112, p. 179, Fig. 113, p. 180, Fig. 114, p. 181, Fig. 115, A, B, p. 184, measurements & meristics Table 30, p. 172, Tables 31, 32, p. 173, Tables 33, 34, p. 177, Tables 35, 36, p. 182, Tables 37, 38, p. 183, Table 39, p. 184. Earth Sciences Museum ESB B-01. Molds made of both part & counterpart by L. Allan Lindoe, 2011. CCIS L2-245, Eocene 07, 04; CCIS L2-245, Eocene 07, 05

Wilson, Mark V. H. 1982. A new species of the fish *Amia* from the Middle Eocene of British Columbia. Palaeontology, Vol. 25(2):413–424.

PARATYPES:

UALVP13801 scales (a & b) & 2 branchial tooth plates. Wilson, 1982, Descr. p. 415. Grande and Bemis, 1998, SVP Memoir 4:1–690. Descr. p. 172. CCIS L2-245, Eocene 07, 03

UALVP13804 right dentary & maxilla. Wilson, 1982, Descr. p. 415, & fig. Fig. 6, C, p. 420. Grande and Bemis, 1998, SVP Memoir 4:1–690. Descr. p. 172. CCIS L2-245, Eocene 07, 03

UALVP13805 right extrascapular, Right 4th infraorbital, Left 5th infraorbital. Wilson, 1982, Descr. p. 415, & fig. Fig. 6, F, p. 420. Grande and Bemis, 1998, SVP Memoir 4:1–690. Descr. p. 172. CCIS L2-245, Eocene 07, 02

UALVP13806 branchial tooth plate. Wilson, 1982, Descr. p. 415, & fig. Fig. 6, D, p. 420. Grande and Bemis, 1998, SVP Memoir 4:1–690. Descr. p. 172. CCIS L2-245, Eocene 07, 03

UALVP13812 right opercle (a & b). Wilson, 1982, Descr. p. 415, & fig. Fig. 6, E, p. 420. Grande and Bemis, 1998, SVP Memoir 4:1–690. Descr. p. 172. Incorrectly described as specimen UAVP 138126 by Grande and Bemis, Table 31, p. 173. CCIS L2-245, Eocene 07, 02

Osteichthyes Huxley, 1880
   Holostei Müller, 1845
      Halecomorphi Cope, 1872
         Ophiopsidae Bartram, 1975
UALVP53662 Cast of HOLOTYPE Cast of IGM3460 original at IGM Instituto de Geología, Universidad Nacional Autónoma de México *Teoichthys kallistos* Applegate, 1988
Collector: Benjamín Aranguthy, 1982
Mexico: Puebla, Tepexi de Rodriguez, Zone 2 of the Aranguthy quarry, IGM-locality 370, which is part of the Tlayua quarry. Cretaceous-Cretaceous Early-Albian, Tlayúa Formation. Whole

fish. Applegate, 1992, Descr. pp. 200–204, & fig. Fig. 2, p. 202, Fig. 3, p. 203. CCIS L2-245, Cretaceous 7, 1

Applegate, Shelton Pleasants 1988. A new genus and species of a holostean belonging to the family Ophiopsidae, *Teoichthys kallistos*, from the Cretaceous near Tepexi de Rodrigquez. Puebla. Universidad Naciojnal Autonoma de Mexico, Insituto de Geologia, UNAM, Revista, Vol. 7(2): 200–205

Osteichthyes Huxley, 1880
  Acanthomorpha Rosen, 1973 (sensu Johnson and Patterson 1993)
    Polymixiiformes Lowe, 1838
      Family *incertae sedis*
UALVP56113   HOLOTYPE *Cumbaaichthys oxyrhynchus* **Murray, 2016**
Collectors: Murray, Alison M.; Holmes, Robert B.; Cumbaa, Stephen L.; and Day, Richard G., August 2010.
Canada: Northwest Territories, Lac des Bois, Site: 779. 66°52′N. Lat., 125°22′W. Long. Cretaceous-Cretaceous Late-Turonian. An almost complete fish, missing only the posteriormost tip of the fin rays of the dorsal lobe of the caudal fin, with a counterpart for the anterior portion. Murray, 2016, Descr. pp. 105–107, & fig. Fig. 1, A, B, C, p. 103, Fig. 2, p. 104, Fig. 3, p. 106, Fig. 4, p. 107. CCIS L2-245, Cretaceous 11, 2.

Murray, Alison M. 2016. Mid-Cretaceous acanthomorph fishes with the description of a new species from the Turonian of Lac des Bois, Northwest Territories, Canada. Vertebrate Anatomy Morphology Palaeontology 1(1):101–115

Osteichthyes Huxley, 1880
  Polymixiformes Lowe, 1838
    Boreiohydriidae Murray and Cumbaa, 2013
UALVP54046 (UALVP54046.1 part, UALVP54046.2 counterpart) HOLOTYPE *Boreiohydrias dayi* **Murray and Cumbaa, 2013**

Collector: Cumbaa, Stephen L.; Murray, Alison M.; Holmes, Robert B.; and Day, Richard G., August 2010. Canada: Northwest Territories, Lac des Bois, Site: 779. 66°52'N. Lat., 125°22'W. Long. Cretaceous-Cretaceous Late-Turonian early. Complete fish preserved in part & counterpart, Descr. Murray and Cumbaa 2013, Descr. pp. 294–296, & fig. Fig. 1, a, b, p. 295, Fig. 2, p. 296, Fig. 3, a, b, p. 297, Fig. 4, p. 297. CCIS: L2-245: Cretaceous 11, 2

Murray, Alison M., and Cumbaa, Stephen L. 2013. Early Turonian Acanthomorphs from Lac Des Bois, Northwest Territories, Canada. Journal of Vertebrate Paleontology, Vol. 33(2): 293–300

Osteichthyes Huxley, 1880
  Perleidiformes, Berg, 1937
    Platysiagidae Brough, 1939
UALVP19119 HOLOTYPE *Helmolepis cyphognathus* **Neuman and Mutter, 2005**
Collectors: Lindoe, L. Allan; Wilson, Mark V. H., 1983
Canada: British Columbia, Wapiti Lake, Cirque F. 54°32'N. Lat., 140°45.4'W. Long. Triassic-Triassic Early-Smithian, Sulphur Mountain. Two partial, part & counterpart, including two peels. Neuman and Mutter, 2005, Descr. pp. 28–33, Fig. 2, A, B, C, p. 28. Fig. 6, p. 31, Fig. 8, Fig. 9, p. 32. CCIS L2-245, Triassic 01, 07

Neuman, Andrew G., and Mutter, Raoul J. 2005. *Helmolepis cyphognathus*, sp. nov., a new platysiagid actinopterygian from the Lower Triassic Sulphur Mountain Formation (British Columbia, Canada). Canadian Journal of Earth Sciences, Vol. 42(1):25–36
PARATYPES:
UALVP1304 remains. Neuman and Mutter, 2005, Descr. p. 28. CCIS L2-245, Triassic 01, 05
UALVP1310 skeleton. Neuman and Mutter, 2005, Descr. p. 28. P. 31. CCIS L2-245, Triassic 01, 05
UALVP1317 skeleton. Neuman and Mutter, 2005, Descr. p. 28. CCIS L2-245, Triassic 01, 05

UALVP1319 partial skeleton. Neuman and Mutter, 2005, Descr. p. 28. CCIS L2-245, Triassic 01, 05

UALVP17122 (?) incorrectly cited as catalogue number UALVP17122. Scale(?). Neuman and Mutter, 2005, Descr. p. 28. UALVP17122 in catalogue is from Canada: Alberta, Joffre, Joffre Bridge NE#1 river cut. Tertiary-Paleocene, Paskapoo

UALVP18988 partial. Neuman and Mutter, 2005, Descr. p. 28

UALVP18991 partial. Neuman and Mutter, 2005, Descr. p. 28. CCIS L2-245, Triassic 01, 05

UALVP19011 scales. Neuman and Mutter, 2005, Descr. p. 28. CCIS L2-245, Triassic 01, 05

UALVP19015 partial. Neuman and Mutter, 2005, Descr. p. 28. CCIS L2-245, Triassic 01, 05

UALVP19016 nine specimens. Neuman and Mutter, 2005, Descr. p. 28. CCIS L2-245, Triassic 01, 07

UALVP19024 two partial, part & counterpart. Neuman and Mutter, 2005, Descr. p. 28. CCIS L2-245, Triassic 01, 05

UALVP19026 partial. Neuman and Mutter, 2005, Descr. p. 28. CCIS L2-245, Triassic 01, 07

UALVP19059 (peel TMP 91.3.3) complete; counterpart UALVP19097. Neuman and Mutter, 2005, Descr. p. 28. CCIS L2-245, Triassic 01, 07

UALVP19071 body. Neuman and Mutter, 2005, Descr. p. 28. CCIS L2-245, Triassic 01, 05

UALVP19094 complete. Neuman and Mutter, 2005, Descr. p. 28. CCIS L2-245, Triassic 01, 05

UALVP19097 complete; counterpart UALVP19059. Neuman and Mutter, 2005, Descr. p. 28. CCIS L2-245, Triassic 01, 07

UALVP19100 complete; Neuman and Mutter, 2005, Descr. p. 28. CCIS L2-245, Triassic 01, 05

UALVP19120 two complete, part & counterpart. Neuman and Mutter, 2005, Descr. p. 28. CCIS L2-245, Triassic 01, 05

UALVP19122 partial skull & anterior body. Neuman and Mutter, 2005, Descr. p. 28. CCIS L2-245, Triassic 01, 05

UALVP22539 two specimens complete. Neuman and Mutter, 2005, Descr. p. 28. CCIS L2-245, Triassic 01, 05

UALVP22540 fish. Neuman and Mutter, 2005, Descr. p. 28. CCIS L2-245, Triassic 01, 05

UALVP22541 fish head & anterior body. Neuman and Mutter, 2005,Descr. p. 28. CCIS L2-245, Triassic 01, 05

UALVP22543 fish. Neuman and Mutter, 2005, Descr. p. 28. CCIS L2-245, Triassic 01, 05

UALVP22544 fish head. Neuman and Mutter, 2005, Descr. p. 28. CCIS L2-245, Triassic 01, 05

UALVP22545 fish. Neuman and Mutter, 2005, Descr. p. 28. CCIS L2-245, Triassic 01, 05

UALVP22546 fish head & anterior body. Neuman and Mutter, 2005, Descr. p. 28, Single supraorbital present p. 29, p. 33. CCIS L2-245, Triassic 01, 05

UALVP22547 lacking head. Neuman and Mutter, 2005, Descr. p. 28. CCIS L2-245, Triassic 01, 05

UALVP22548 lacking tail. Neuman and Mutter, 2005, Descr. p. 28. CCIS L2-245, Triassic 01, 05

UALVP22549 fish. Neuman and Mutter, 2005, Descr. p. 28. CCIS L2-245, Triassic 01, 05

UALVP22550 partial. Neuman and Mutter, 2005, Descr. p. 28. CCIS L2-245, Triassic 01, 05

UALVP22551 (peel TMP 91.3.1) lacking tail. Neuman and Mutter, 2005, Descr. p. 28, p. 31, p. 32. CCIS L2-245, Triassic 01, 07

UALVP22552 partial. Neuman and Mutter, 2005, Descr. p. 28. CCIS L2-245, Triassic 01, 05

UALVP29713 Neuman and Mutter, 2005, Descr. p. 28. CCIS L2-245, Triassic 01, 05

UALVP29714 Neuman and Mutter, 2005, Descr. p. 28. CCIS L2-245, Triassic 01, 05

UALVP29715 Neuman and Mutter, 2005, Descr. p. 28. CCIS L2-245, Triassic 01, 05

UALVP29716 Neuman and Mutter, 2005, Descr. p. 28. CCIS L2-245, Triassic 01, 05

UALVP29717 Neuman and Mutter, 2005, Descr. p. 28. CCIS L2-245, Triassic 01, 05

UALVP29718 Neuman and Mutter, 2005, Descr. p. 28. CCIS L2-245, Triassic 01, 05

UALVP29720 two specimens. Neuman and Mutter, 2005, Descr. p. 28 as paratypes, but on the same page also listed as specimens tentatively described this species. CCIS L2-245, Triassic 01, 05

UALVP31624 peels of TMP 83.206.85. Neuman and Mutter, 2005, Descr. p. 28. CCIS L2-245, Triassic 01, 15

UALVP31625 peel of TMP 83.205.72, skull of fish. Neuman and Mutter, 2005, Descr. p. 28. CCIS L2-245, Triassic 01, 15

UALVP31627 peel of TMP 83.205.76, skull of fish. Neuman and Mutter, 2005, Descr. p. 28. CCIS L2-245, Triassic 01, 15

UALVP46652 head, part & counterpart. Neuman and Mutter, 2005, Descr. p. 28, p. 29. CCIS L2-245, Triassic 07, 01

UALVP46653 head, body, fins missing. Neuman and Mutter, 2005, Descr. p. 28. CCIS L2-245, Triassic 07, 01

UALVP46654 head, body, fins missing; Cirque Fs. Neuman and Mutter, 2005, Descr. p. 28. CCIS L2-245, Triassic 07, 01

UALVP46655 head & anterior body, three parts. Neuman and Mutter, 2005, Descr. p. 28. CCIS L2-245, Triassic 07, 01

UALVP46656 head & anterior body, scattered; Neuman and Mutter, 2005, Descr. p. 28. CCIS L2-245, Triassic 07, 01

UALVP46657 head & anterior body; Neuman and Mutter, 2005, Descr. p. 28. CCIS L2-245, Triassic 07, 01

UALVP46660 nearly complete. Neuman and Mutter, 2005, Descr. p. 28, Mutter, 2004. CCIS L2-245, Triassic 08, 07

UALVP46671 head & **UALVP46671-T1** (thin section). Neuman and Mutter, 2005, Descr. p. 28

UALVP46672 nearly complete. Neuman and Mutter, 2005, Descr. p. 28. CCIS L2-245, Triassic 06, 16

UALVP46673 nearly complete, weathered, 3 supraorbitals present Cirque C. Neuman and Mutter, 2005, Descr. p. 28, p. 29. CCIS L2-245, Triassic 06, 16

UALVP46674 nearly complete. Neuman and Mutter, 2005, Descr. p. 28. CCIS L2-245, Triassic 06, 16

UALVP46676 head & anterior body, weathered. Neuman and Mutter, 2005, Descr. p. 28. CCIS L2-245, Triassic 06, 16

UALVP46677 head & anterior body, Cirque Fs., Neuman and Mutter, 2005, Descr. p. 28. CCIS L2-245, Triassic 06, 16

UALVP46678 head & anterior body, weathered, distorted; Cirque Cs-t. Neuman and Mutter, 2005, Descr. p. 28. CCIS L2-245, Triassic 06, 16

UALVP46679 head & anterior body, weathered; **UALVP46679-T1** and **UALVP46679-T2** thin sections. Neuman and Mutter, 2005, Descr. p. 28, & fig. Fig. 7, p. 31

UALVP46680 nearly complete, weathered. Neuman and Mutter, 2005, Descr. p. 28, p. 29. CCIS L2-245, Triassic 06, 16

UALVP46681 nearly complete, weathered. Neuman and Mutter, 2005, Descr. p. 28, p. 29. CCIS L2-245, Triassic 07, 02

UALVP46682 nearly complete, weathered. Neuman and Mutter, 2005, Descr. p. 28. CCIS L2-245, Triassic 07, 02

UALVP46683 unprepared specimen. Neuman and Mutter, 2005, Descr. p. 28. CCIS L2-245, Triassic 07, 02

UALVP46731 several specimens, partly preserved. Neuman and Mutter, 2005, Descr. p. 28. CCIS L2-245, Triassic 07, 08; CCIS L2-245, Triassic 08, 13

UALVP46782 partial head. T corridor. Neuman and Mutter, 2005, Descr. p. 28. CCIS L2-245, Triassic 05, 14

Osteichthyes Huxley, 1880
  Holostei Müller, 1845
    Halecomorphi Cope, 1872
      Macrosemiiformes Grande and Bemis, 1998
      Macrosemiidae Thiollière, 1858
UALVP47133 HOLOTYPE *Agoultichthys chattertoni* **Murray and Wilson, 2009**
Collector: Chatterton, Brian D. E., and Gibb, Stacey, 2006

Morocco, Agoult. Cretaceous-Cretaceous Late-Cenomanian, Akrabou Formation. Whole fish. Murray and Wilson, 2009, Descr. pp. 431–435, & fig. Fig. 2, 3, p. 431; Fig. 4–5, p. 432; Fig. 6–7, p. 433. Note: UALVP43599 incorrectly identified as Holotype in Murray *et al.* 2013, p. 530, it is actually not a type as there was only one specimen, UALVP47133, in the original description.

Murray, A. M., and Wilson, M. V. H. 2009. A new Late Cretaceous Macrosemiid fish (Neopterygii, Halecostomi) from Morocco, with temporal and geographical range extensions for the family. Palaeontology, Vol. 52(2):429–440.

Murray, Alison M., Wilson, Mark V. H., Gibb, Stacey, and Chatterton, Brian D. E. 2013. Additions to the Late Cretaceous (Cenomanian/Turonian) actinopterygian fauna from the Agoult locality, Akrabou Formation, Morocco, and comments on the palaeoenvironment. pp. 525–548. *IN:* Arratia, G., and Schultze, H.-P. (editors) Mesozoic Fishes 5-Global Diversity and Evolution. Friedrich Pfeil. Munchen, Germany, 560 pp.

Osteichthyes Huxley, 1880
    Osteoglossiformes Regan, 1909
        Osteoglossoidei Regan, 1909
            Osteoglossidae Bonaparte, 1850
                Heterotidinae Nelson, 1968
UALVP23705 a, b HOLOTYPE *Joffrichthys symmetropterus* Li and Wilson, 1996
Collector: Lindoe, L. Allan, 1987
Canada: Alberta, Joffre, Joffre Bridge SW Fish Layer. 52°16.1′N. Lat., 113°35.7′W. Long. Tertiary-Paleocene, Paskapoo Formation. Nearly complete individual in part and counterpart (UALVP 23705a, 23705b). Li and Wilson, 1996, Descr. measurements, p. 200, meristics, p. 201, hyomandibular p. 203, & fig. Fig. 1, p. 200, Fig. 3, p. 202, Fig. 5, p. 203, Murray and Wilson, 2005, Zool. J. Linn. Soc. 144:213–228, Descr. p. 214. Wilson and Murray, 2008,

Fishes & the Break-Up of Pangaea, 295:185–219, Fig. 4, b, p. 189. CCIS L2-245, Paleocene 05, 01

Li, Guo-qing, and Wilson, M. V. H. 1996. The discovery of Heterotidinae (Teleostei: Osteoglossidae) from the Paleocene Paskapoo formation of Alberta, Canada. Journal of Vertebrate Paleontology, Vol. 16(2):198–209.

PARATYPES:

UALVP15069 left dentary, Wilson, 1980, CJES, Vol. 17(3):fig. Fig. 1B, p. 308. Li and Wilson, 1996, Descr. p. 200, p. 203, & fig. Fig. 4, C, p. 202. Paleocene 3, 19

UALVP15995 fish bone fragments, impression of frontal. Li and Wilson, 1996, Descr. p. 200, & fig. Fig. 4, A, p. 202

UALVP17024 jaw fragment with teeth, dentary. Li and Wilson, 1996, Descr. p. 200, p. 203

UALVP17068 jaw fragment with teeth, maxilla. Li and Wilson, 1996, Descr. p. 200, p. 202

UALVP17241 jaw fragments with teeth, parasphenoid. Li and Wilson, 1996, Descr. p. 200, p. 202

UALVP17242 large bone fragments, preopercle. Li and Wilson, 1996, Descr. p. 200, & fig. Fig. 4, E, p. 202. Paleocene 3, 19

UALVP17317 hiodontid bone fragments & scales, maxilla. Li and Wilson, 1996, Descr. p. 200

UALVP17327 hiodontid scales, bone fragments, jaw fragments with teeth, maxilla, a tooth plate of the basihyal, parasphenoid. Li and Wilson, 1996, Descr. p. 200, p. 202, p. 203. Paleocene 3, 19

UALVP17846 opercles. Li and Wilson, 1996, Descr. p. 200, p. 202

UALVP17847 opercles. Li and Wilson, 1996, Descr. p. 200, p. 202

UALVP17848 opercles. Li and Wilson, 1996, Descr. p. 200, p. 202, & fig. Fig. 4, B, p. 202

UALVP17852 opercles. Li and Wilson, 1996, Descr. p. 200, p. 202

UALVP17853 dentary. Li and Wilson, 1996, Descr. p. 200, p. 203

UALVP17854 dentary. Li and Wilson, 1996, Descr. p. 200, p. 203

UALVP31545 fish, complete, part & counterpart. Li and Wilson, 1996, Descr. measurements p. 200, p. 201, basihyal p. 203, & fig

Fig. 2, A, p. 201. Murray and Wilson, 2005, Zool. J. Linn. Soc. 144:213–228, Descr. p. 214. CCIS L2-245, Paleocene 05, 01
UALVP37128 part & counterpart articulated specimen. Li and Wilson, 1996, Descr. measurements and meristics, p. 201, retroarticular, p. 203, caudal fin lower lobe, p. 204. & fig. Fig. 2, B, p. 201, Fig. 4, D, p. 202
UALVP37136 maxilla. Li and Wilson, 1996, Descr. p. 200, p. 202. CCIS L2-245, Paleocene 05, 01

Osteichthyes Huxley, 1880
    Osteoglossiformes Regan, 1909
        Hiodontidae Valenciennes, 1846
UALVP13405 Cast of HOLOTYPE Originals at National Museum of Canada, casts of NMC2156 a, b, *Eohiodon rosei* **(Hussakof, 1916)**
Collector: Rose, B., 1906.
Canada: British Columbia, Kamloops Lake, Red Point, north shore of Kamloops Lake. Eocene. Whole fish. Hussakof, 1916, Descr. pp. 18–20, fig. Fig. 1. Placed in *Eohiodon* by Cavender, 1966, pp. 311–320

    Hussakof, L. 1916. A new cyprinid fish, *Leuciscus rosei*, from the Miocene of British Columbia. American Journal of Science, Vol. 42, ser. 4, article 2:18–20

    Cavender, T. M. 1966. Systematic position of the North American Eocene fish, "*Leuciscus*" *rosei* Husakof. Copeia 1966(2):311–320.

Osteichthyes Huxley, 1880
    Osteoglossiformes Regan, 1909
        Hiodontidae Valenciennes, 1846
UALVP13227 HOLOTYPE *Eohiodon woodruffi* **Wilson, 1978**
Collectors: Lindoe, L. Allan, and Wilson, Mark V. H., 1976
The United States: Washington, Republic, Tom Thumb Mine, 48°42′N. Lat., 118°45′W. Long. Tertiary-Eocene-Bridgerian NALMA, Klondike Mountain Formation, Tom Thiumb Tuff

Member. Complete, part & counterpart. Wilson, 1978, Descr. p. 681, & fig. Fig. 2A, p. 682. Hilton and Grande, 2008, Descr. . p. 224

Wilson, Mark V. H. 1978. *Eohiodon woodruffi* n. sp. (Teleostei, Hiodontidae), from the Middle Eocene Klondike Mountain Formation near Republic, Washington. Canadian Journal of Earth Sciences, Vol. 15(5):679–686.

PARATYPES:

UALVP13225 lacking head. Wilson, 1978, Descr. p. 681, p. 683, p. 684. CCIS L2-245, Eocene 07

UALVP13241 near complete. Wilson, 1978, Descr. p. 681, Table 1, p. 683

UALVP13244 Wilson, 1978, Descr. p. 681, Table 1, p. 683

UALVP13250 trunk. Wilson, 1978, Descr. p. 681, Table 1, p. 683

UALVP13262 partial. Wilson, 1978, Descr. p. 681, Table 1, p. 683

UALVP13265 near complete. Wilson, 1978, Descr. p. 683, p. 684, & Fig. 2A, p. 682. Hilton and Grande, 2008, Descr. pp. 224, 237

Osteichthyes Huxley, 1880
   Osteoglossiformes Regan, 1909
      Hiodontidae Valenciennes, 1846
UALVP40729 Cast of HOLOTYPE Cast of FMNH PF 10424, formerly UMVP 6499, University of Minnesota, Vertebrate Paleontology Collections. *Eohiodon falcatus* **Grande, 1979 (= now** *Eohiodon woodruffi* **Wilson, 1978)**
Collector: Lance Grande, 1977
USA: Wyoming: Lincoln County: about 16 km. NW of Kemmerer, Thompson Ranch, Tynsky Quarry, NW1/4, SW1/4, Sec. 22, T22N, R117W.
Early Eocene: Green River Fm.: Fossil Butte Member Whole fish. Grande, 1979, Descr. pp. 103–111 & fig. Fig. 1, p. 104. CCIS L2-245, Eocene 14, 16
See Bruner, 1992, Fieldiana, p. 46.

Grande, Lance. 1979. *Eohiodon falcatus*, a new species of hiodontid (Pisces) from the Late Early Eocene Green River Formation of Wyoming. Journal of Paleontology, Vol. 53(1):103–111.

Osteichthyes Huxley, 1880
  Osteoglossiformes Regan, 1909
    Hiodontidae Valenciennes, 1846
UALVP38875 HOLOTYPE *Hiodon consteniorum* **Li and Wilson, 1994, p. 155**
Collectors: Constenius, J. Norm, and Constenius, Leona
The United States: Montana, Disbrow Creek, Tertiary-Eocene-Eocene early to middle, Kishenehn Formation [formerly thought to be Oligocene, Rupelian]. Nearly complete; well-preserved; counterpart returned to Constenius. Li and Wilson, 1994, p. 155, & fig. Fig. 1, p. 156; Fig. 2, p. 157; Fig. 3, A, p. 159; Fig. 6, C, p. 162. Hilton and Grande, 2008, Descr. p. 224. Wilson and Murray, 2008, Fishes & the Break-Up of Pangaea, fig. Fig. 2, d, p. 187. CCIS L2-245, Eocene 18, 16
  Li, G.-Q., and Wilson, M. V. H. 1994. An Eocene species of *Hiodon* from Montana, its phylogenetic relationships, and the evolution of the postcranial skeleton in the Hiodontidae (Teleostei). Journal of Vertebrate Paleontology, Vol. 14(2):153–167.
PARATYPE:
UALVP24200 almost complete. Li and Wilson, 1994, pp. 155, 158, & fig. Fig. 1, B, p. 156; Fig. 3, B, p. 159. Hilton and Grande, 2008, Descr. pp. 224, 226. CCIS L2-245, Eocene 18, 16

Osteichthyes Huxley, 1880
  Clupeomorpha Greenwood, Rosen, Weitzman, and Myers, 1966
    Ellimmichthyiformes Grande, 1982
      Scutatuspinosidae Silva Santos and Silva Corréa, 1985
UALVP17535 HOLOTYPE *Foreyclupea loonensis* **Vernygora, Murray, and Wilson, 2016**
Collector: Fox, Richard C., 1973
Canada: Nunavut, Cameron Hills, Hay River #1. Cretaceous-Cretaceous Early-Albian, Loon River Formation. Partial. Hermus *et al.,* 2004, Mes. Fishes 3, Descr. pp. 449–461, p. 452 Vernygora *et al.,* 2016, Descr. p. 332, & fig. Fig. 1 a, b, p. 333, Fig. 3 a, b, p. 335, measurements Table 1, p. 336. CCIS L2-245, Cretaceous 02, 11

Vernygora, Oksana, Murray, Alison M., and Wilson, Mark V. H. 2016. A primitive clupeomorph from the Albian Loon River Formation (Northwest Territories, Canada). Canadian Journal of Earth Sciences, Vol. 53: 331–342.

Osteichthyes Huxley, 1880
Clupeomorpha Greenwood, Rosen, Weitzman, and Myers, 1966
Ellimmichthyiformes Grande, 1982
Sorbinichthyidae Bannikov and Bacchia, 2000
UALVP51640 HOLOTYPE *Sorbinichthyes africanus* Murray and Wilson, 2011
Morocco: Ouarzazate, Agoult, site KK0809. Cretaceous-Cretaceous Late-Cenomanian, Akrabou Formation. Whole fish. Murray and Wilson, 2011, fig. Fig. 1, p. 2, Fig. 4, p. 5, Table 1, p. 8. Murray *et al.* 2013, Descr. p. 532
Murray, Alison M., and Wilson, Mark V. H. 2011. A new species of *Sorbinichthys* (Teleostei: Clupeomorpha: Ellimmichthyiformes) from the Late Cretaceous of Morocco. Canadian Journal of Earth Sciences, Vol. 48:1–9.
PARATYPES:
UALVP47186 whole fish, part & counterpart. Murray and Wilson, 2011, fig. Fig. 2, B, p. 3, Table 1, p. 8. Murray *et al.*, 2013, Descr. p. 532
UALVP51641 whole fish, part & counterpart. Murray and Wilson, 2011, fig. Fig. 2, A, p. 3, Fig. 5, p. 6, Fig. 6, p. 7, Table 1, p. 8. Murray *et al.*, 2013, Descr. p. 532

Osteichthyes Huxley, 1880
Ellimmichthyiformes Grande, 1982
Paraclupeidae Chang and Chou, 1977
UALVP47178 HOLOTYPE *Thorectichthys marocensis* Murray and Wilson, 2013
Collector: Chatterton, Brian D. E., and Gibb, Stacey, 2006
Morocco, Agoult. Cretaceous-Cretaceous Late-Cenomanian. Whole fish, part & counterpart. Murray and Wilson, 2013, Descr.

pp. 269–276, & fig. Fig. 1, A, B, E, p. 270; Fig. 2, p. 272; Murray *et al.*, 2013, Descr. p. 534

Murray, Alison M., and Wilson, Mark V. H. 2013. Two new paraclupeid fishes (Clupeomorpha: Ellimmichthyiformes) from the Upper Cretaceous of Morocco. pp. 267–290 *IN:* Arratia, G., and Schultze, H.-P. (editors). Mesozoic Fishes 5-Global diversity and Evolution Dr. Friedrich Pfeil Munchen, Germany. 560 pp.

<u>PARATYPES:</u>

UALVP47134 whole fish, part & counterpart. Murray and Wilson, 2013, Fig. 5, p. 275; Descr. pp. 270, 275–276. Murray *et al.*, 2013, Mes. Fishes 5, Descr. p. 534

UALVP51612 whole fish, "20." Murray and Wilson, 2013, Mes. Fishes 5, Descr. pp. 267–290 Table 1, p. 271

UALVP51647 whole fish, part & counterpart. Murray and Wilson, 2013, Descr. p. 270, p. 274; Fig. 4, p. 274. Murray and Wilson, 2013. Murray *et al.*, 2013, Mes. Fishes 5, Descr. p. 534

UALVP51648 whole fish. Murray and Wilson, 2013, Descr. Table 1, p. 271. Murray *et al.*, 2013, Mes. Fishes 5, Descr. p. 534

UALVP51649 3 fish, part & counterpart, 1 Paraclupeidae +2 Clupeiformes. Murray and Wilson, 2013, Descr. p. 270, p. 274. Murray *et al.*, 2013, Mes. Fishes 5, Descr. p. 534

UALVP51650 whole fish, part and counterpart. Murray and Wilson, 2013, fig. Fig. 3, p. 273

UALVP51651 whole fish, part and counterpart. Murray and Wilson, 2013, Mes. Fishes 5, Descr. pp. 267–290 Table 1, p. 271

UALVP51657 whole fish. Murray and Wilson, 2013, Mes. Fishes 5, Descr. p. 270. Murray *et al.*, 2013, Mes. Fishes 5, Descr. p. 534

UALVP51659 whole fish, [all on one slab=UALVP51659, UALVP51660, UALVP51661, on opposite side of slab UALVP51662, UALVP51663]. Murray and Wilson, 2013, Mes. Fishes 5, Descr. p. 270, Table 1, p. 271. Murray *et al.*, 2013, Mes. Fishes 5, Descr. p. 534

UALVP51664 whole fish, and one crustacean. Murray and Wilson, 2013, Mes. Fishes 5, fig. Fig. 1D, P. 270, Descr. pp. 276, 277, Table 1, p. 277, p. 278. Murray *et al.*, 2013, Descr. p. 534.

UALVP51715 1 whole fish, part & counterpart, KK89-M1. Murray and Wilson, 2013, Mes. Fishes 5, Descr. p. 276 & Table 2, p. 277. Murray *et al.*, 2013, Mes. Fishes 5, Descr. p. 534 & fig. Fig. 9, p. 535

Osteichthyes Huxley, 1880
    Ellimmichthyiformes Grande, 1982
        Paraclupeidae Chang and Chou, 1977
UALVP51653 HOLOTYPE *Thorectichthys rhadinus* **Murray and Wilson, 2013**
Morocco: Ouarzazate, Agoult, site KK0809. Cretaceous-Cretaceous Late-Cenomanian, Akrabou Formation. Whole fish. Murray and Wilson, 2013, fig. Fig. 1C, p. 270, Descr. p. 276, Table 2, p. 277, p, 277. Murray *et al.*, 2013, Mes. Fishes 5, Descr. p. 534
    Murray, Alison M., and Wilson, Mark V. H. 2013. Two new paraclupeid fishes (Clupeomorpha: Ellimmichthyiformes) from the Upper Cretaceous of Morocco. pp. 267–290 *IN:* Arratia, G., and Schultze, H.-P. (editors) Mesozoic Fishes 5-Global diversity and Evolution Dr. Friedrich Pfeil Munchen, Germany. 560 pp.
PARATYPES:
UALVP51664 whole fish, and 1 crustacean. Murray and Wilson, 2013, Mes. Fishes 5, fig. Fig. 1D, P. 270; Descr. pp. 276, 277, Table 1, p. 277, p. 278. Murray *et al.*, 2013, Descr. p. 534
UALVP51715 1 whole fish, part & counterpart. Murray and Wilson, 2013, Mes. Fishes 5, Descr. p. 276 & Table 2, p. 277. Murray *et al.*, 2013, Descr. p. 534 & fig. Fig. 9, p. 535

Osteichthyes Huxley, 1880
    Division Teleostei (sensu Nelson, 1969)
        Clupeomorpha Greenwood, Rosen *et al.* 1966
            Clupeiformes (sensu Greenwood, Rosen *et al.* 1966)
            Family *incertae sedis*
UALVP8606 HOLOTYPE *Erichalcis arcta* **Forey, 1975, p. 153**
Collector: Lindoe, L. Allan, 1972
Canada: Nunavut, Cameron Hills, Hay River #1, 60°01′N. Lat., 116°57′W. Long., bed of Hay River. Cretaceous-Cretaceous

Early-Albian, Loon River Formation. Complete fish. Forey, 1975, Descr. p. 153 & fig. Fig. 2, A, p. 155. Hermus *et al.*, 2004. Mes. Fishes 3, Descr. pp. 449–461, p. 449, p. 452, fig. Fig. 5, p. 455. Arratia, 2013 SVP Memoir 13: p. 12. CCIS L2-245, Cretaceous 02, 11

Forey, Peter L. 1975. A fossil clupeomorph fish from the Albian of the Northwest Territories of Canada, with notes on cladistic relationships of clupeomorphs. Journal of Zoology, Vol. 175(2):151–177

PARATYPES:

UALVP8589 partial fish (part A & counterpart B). Forey, 1975, Descr. p. 153, & fig. Fig. 7, b, p. 159. Hermus *et al.*, 2004 Mes. Fishes 3, Descr. p. 452. CCIS L2-245, Cretaceous 02, 11

UALVP8590 jaws. Forey, 1975, Descr. p. 153. Hermus *et al.*, 2004, Mes. Fishes 3, Descr. p. 452. CCIS L2-245, Cretaceous 02, 11

UALVP8591 jaws [**specimen listed as missing April 1992**]. Forey, 1975, Descr. p. 153, Hermus *et al.*, 2004, Mes. Fishes 3, Descr. p. 452. CCIS L2-245, Cretaceous 02, 08

UALVP8592 jaws. Forey, 1975, Descr. p. 153, & fig. Fig. 6, A, B, p. 158. Hermus *et al.*, 2004, Mes. Fishes 3, Descr. p. 452. CCIS L2-245, Cretaceous 02, 08

UALVP8593 jaws. Forey, 1975, Descr. p. 153, & fig. Fig. 6, A, B, p. 158. Hermus *et al.*, 2004, Mes. Fishes 3, Descr. p. 452. CCIS L2-245, Cretaceous 02, 11

UALVP8594 jaws. Forey, 1975, Descr. p. 153. Hermus *et al.*, 2004, Mes. Fishes 3, Descr. p. 452. CCIS L2-245, Cretaceous 02, 08

UALVP8595 jaws. Forey, 1975, Descr. p. 153. Hermus *et al.*, 2004, Mes. Fishes 3, Descr. p. 452. CCIS L2-245, Cretaceous 02, 11

UALVP8596 skull roof, hyoid bar & snout. Forey, 1975, Descr. p. 153, & fig. Fig. 3, A, B, p. 156. Hermus *et al.*, 2004, Mes. Fishes 3, Descr. p. 452. CCIS L2-245, Cretaceous 02, 11

UALVP8597 tail. Forey, 1975, Descr. p. 153, & fig. Fig. 10, C, p. 163. Hermus *et al.*, 2004, Mes. Fishes 3, Descr. p. 452. CCIS L2-245, Cretaceous 02, 08

UALVP8598 tail. Forey, 1975, Descr. p. 153, & fig. Fig. 10, A, p. 163. Arratia, 2013 SVP Memoir 13: p. 12. Hermus *et al.,* 2004, Mes. Fishes 3, Descr. p. 452. CCIS L2-245, Cretaceous 02, 11

UALVP8599 head & anterior trunk a, b. Forey, 1975, Descr. p. 153. Hermus *et al.,* 2004, Mes. Fishes 3, Descr. p. 452. CCIS L2-245, Cretaceous 02, 11

UALVP8600 head. Forey, 1975, Descr. p. 153, & fig. Fig. 2, B, p. 155. Hermus *et al.,* 2004, Mes. Fishes 3, Descr. p. 452. CCIS L2-245, Cretaceous 02, 11

UALVP8601 head [**specimen listed as missing April 1992**]. Forey, 1975, Descr. p. 153

UALVP8602 head. Forey, 1975, Descr. p. 153; Arratia, 2013 SVP Memoir 13: p. 12. Hermus *et al.,* 2004, Mes. Fishes 3, Descr. p. 452

UALVP8603 jaws. Forey, 1975, Descr. p. 153. Hermus *et al.,* 2004, Mes. Fishes 3, Descr. p. 452. CCIS L2-245, Cretaceous 02, 08

UALVP8604 skull, anterior trunk & pelvic girdle. Forey, 1975, Descr. p. 153. Hermus *et al.,* 2004, Mes. Fishes 3, Descr. p. 452

UALVP8605 snout & jaws. Forey, 1975, Descr. p. 153. Hermus *et al.,* 2004, Mes. Fishes 3, pp. Descr. p. 452. CCIS L2-245, Cretaceous 02, 11

UALVP8607 head, showing fragmentary remains of prootic & pterotic. Forey, 1975, Descr. p. 153, p. 156. Hermus *et al.,* 2004, Mes. Fishes 3, Descr. p. 452. CCIS L2-245, Cretaceous 02, 11

UALVP8608 vertebrae, showing 20 or 21 caudal vertebrae. Forey, 1975, Descr. p. 153, p. 161. Hermus *et al.,* 2004, Mes. Fishes 3, Descr. p. 452. CCIS L2-245, Cretaceous 02, 08

UALVP8609 vertebral column & scutes. Forey, 1975, Descr. p. 153. Hermus *et al.,* 2004, Mes. Fishes 3, Descr. p. 452. CCIS L2-245, Cretaceous 02, 11

UALVP8610 parasphenoid. Forey, 1975, Descr. p. 153, & fig. Fig. 5, A, p. 158. Hermus *et al.,* 2004, Mes. Fishes 3, Descr. p. 452. CCIS L2-245, Cretaceous 02, 08

UALVP8611 tail. Forey, 1975, Descr. p. 153. Hermus *et al.,* 2004, Mes. Fishes 3, Descr. p. 452. CCIS L2-245, Cretaceous 02, 08

UALVP8612 head. Forey, 1975, Descr. p. 153, & fig. Fig. 5, B, p. 158. Hermus *et al.*, 2004, Mes. Fishes 3, Descr. p. 452. Arratia, 2013 SVP Memoir 13: Descr. p. 12

UALVP8613 head. Forey, 1975, Descr. p. 153. Hermus *et al.*, 2004, Mes. Fishes 3, Descr. p. 452. CCIS L2-245, Cretaceous 02, 11

UALVP8614 head. Forey, 1975, Descr. p. 153. [**specimen missing as of April 1992**]

UALVP8615 head. Forey, 1975, Descr. p. 153. [**specimen missing as of April 1992**]

UALVP8616 jaws. Forey, 1975, Descr. p. 153. Hermus *et al.*, 2004, Mes. Fishes 3, Descr. p. 452. CCIS L2-245, Cretaceous 02, 08

UALVP8617 jaws. Forey, 1975, Descr. p. 153. Hermus *et al.*, 2004, Mes. Fishes 3, Descr. p. 452. CCIS L2-245, Cretaceous 02, 11

UALVP8618 jaws. Forey, 1975, Descr. p. 153. Hermus *et al.*, 2004, Mes. Fishes 3, Descr. p. 452. CCIS L2-245, Cretaceous 02, 08

UALVP8619 jaws, quadrate. Forey, 1975, Descr. p. 153. Hermus *et al.*, 2004, Mes. Fishes 3, Descr. p. 452. CCIS L2-245, Cretaceous 02, 11

UALVP8620 jaws. Forey, 1975, Descr. p. 153. Hermus *et al.*, 2004, Mes. Fishes 3, Descr. p. 452. CCIS L2-245, Cretaceous 02, 08

UALVP8621 jaws. Forey, 1975, Descr. p. 153. Hermus *et al.*, 2004, Mes. Fishes 3, Descr. p. 452. CCIS L2-245, Cretaceous 02, 08

UALVP8622 jaws. Forey, 1975, Descr. p. 153. Hermus *et al.*, 2004, Mes. Fishes 3, Descr. p. 452. CCIS L2-245, Cretaceous 02, 08

UALVP8623 jaws Forey, 1975, Descr. p. 153. Hermus *et al.*, 2004, Mes. Fishes 3, Descr. p. 452. CCIS L2-245, Cretaceous 02, 11

UALVP8624 ceratohyal & quadrate. Forey, 1975, Descr. p. 153. Hermus *et al.*, 2004, Mes. Fishes 3, Descr. p. 452. CCIS L2-245, Cretaceous 02, 08

UALVP8625 jaws. Forey, 1975, Descr. p. 153. Hermus *et al.*, 2004, Mes. Fishes 3, Descr. p. 452. CCIS L2-245, Cretaceous 02, 11

UALVP8626 partial head. Forey, 1975, Descr. p. 153. Hermus *et al.*, 2004, Mes. Fishes 3, Descr. p. 452. CCIS L2-245, Cretaceous 02, 08

UALVP8627 head & anterior trunk. Forey, 1975, Descr. p. 153. Hermus *et al.*, 2004, Mes. Fishes 3, Descr. p. 452

UALVP8628 skull & pectoral girdle. Forey, 1975, Descr. p. 153, & fig. Fig. 9. A. B, p. 161. Hermus *et al.*, 2004, Mes. Fishes 3, Descr. p. 452. CCIS L2-245, Cretaceous 02, 11

UALVP8629 imperfect fish. Forey, 1975, Descr. p. 153, p. 162, & fig. Fig. 10. B, p. 163. Hermus *et al.*, 2004, Mes. Fishes 3, Descr. p. 452. CCIS L2-245, Cretaceous 02, 08

UALVP8630 vertebral column & pelvic fins. Forey, 1975, Descr. p. 153. Hermus *et al.*, 2004, Mes. Fishes 3, Descr. p. 452. CCIS L2-245, Cretaceous 02, 11

UALVP8631 tail. Forey, 1975, Descr. p. 153. Hermus *et al.*, 2004, Mes. Fishes 3, Descr. p. 452. CCIS L2-245, Cretaceous 02, 08

UALVP8632 partial. Forey, 1975, Descr. p. 153. Hermus *et al.*, 2004, Mes. Fishes 3, Descr. , p. 452. CCIS L2-245, Cretaceous 02, 08

UALVP8633 pelvic fin. Forey, 1975, Descr. p. 153. Hermus *et al.*, 2004, Mes. Fishes 3, Descr. p. 452. CCIS L2-245, Cretaceous 02, 08

UALVP8634 partial head (specimen & mold). Forey, 1975, Descr. p. 153. Hermus *et al.*, 2004, Mes. Fishes 3, Descr. p. 452. CCIS L2-245, Cretaceous 02, 08

UALVP8635 partial head (specimen & mold). Forey, 1975, Descr. p. 153. Hermus *et al.*, 2004, Mes. Fishes 3, Descr. p. 452. CCIS L2-245, Cretaceous 02, 08

UALVP8636 tail (specimen & mold). Forey, 1975, Descr. p. 153. Hermus *et al.*, 2004, Mes. Fishes 3, Descr. p. 452. CCIS L2-245, Cretaceous 02, 08

UALVP8637 complete fish (specimen & mold). Forey, 1975, Descr. p. 153. Hermus *et al.*, 2004, Mes. Fishes 3, Descr. p. 452. CCIS L2-245, Cretaceous 02, 08

UALVP8638 Forey, 1975, Descr. p. 153. Hermus *et al.*, 2004, Mes. Fishes 3, Descr. p. 452. CCIS L2-245, Cretaceous 02, 08

UALVP8639 pectoral fin & dorsal fin (specimen & mold). Forey, 1975, Descr. p. 153. Hermus *et al.*, 2004, Mes. Fishes 3, Descr. p. 452. CCIS L2-245, Cretaceous 02, 08

UALVP8640 head (specimen & mold). Forey, 1975, Descr. p. 153. Hermus *et al.*, 2004, Mes. Fishes 3, Descr. p. 452. CCIS L2-245, Cretaceous 02, 08

UALVP8641 vertebral column & tail (specimen & mold). Forey, 1975, Descr. p. 153. Hermus *et al.*, 2004, Mes. Fishes 3, Descr. p. 452. CCIS L2-245, Cretaceous 02, 08

UALVP8642 complete fish (specimen & mold). Forey, 1975, Descr. p. 153. Hermus *et al.*, 2004, Mes. Fishes 3, Descr. p. 452. CCIS L2-245, Cretaceous 02, 08

UALVP8643 partial head (specimen & mold). Forey, 1975, Descr. p. 153. Hermus *et al.*, 2004, Mes. Fishes 3, Descr. p. 452. CCIS L2-245, Cretaceous 02, 08

UALVP8644 partial head (specimen & mold). Forey, 1975, Descr. p. 153. Hermus *et al.*, 2004, Mes. Fishes 3, Descr. p. 452. CCIS L2-245, Cretaceous 02, 08

UALVP8699 head & pectoral girdle. Forey, 1975, Descr. p. 153. Hermus *et al.*, 2004, Mes. Fishes 3, Descr. p. 452

UALVP8700 tail impression. Forey, 1975, Descr. p. 153. Hermus *et al.*, 2004, Mes. Fishes 3, Descr. p. 452. CCIS L2-245, Cretaceous 02, 08

UALVP8701 jaws. Forey, 1975, Descr. p. 153

UALVP8702 jaws. Forey, 1975, Descr. p. 153. Hermus *et al.*, 2004, Mes. Fishes 3, Descr. p. 452. CCIS L2-245, Cretaceous 02, 08

UALVP8703 jaws. Forey, 1975, Descr. p. 153. Hermus *et al.*, 2004, Mes. Fishes 3, Descr. p. 452. CCIS L2-245, Cretaceous 02, 08

UALVP8704 jaws. Forey, 1975, Descr. p. 153. Hermus *et al.*, 2004, Mes. Fishes 3, Descr. p. 452. CCIS L2-245, Cretaceous 02, 08

UALVP8705 partial skull. Forey, 1975, Descr. p. 153. Hermus *et al.*, 2004, Mes. Fishes 3, Descr. p. 452. CCIS L2-245, Cretaceous 02, 08

UALVP8706 jaws. Forey, 1975, Descr. p. 153. Hermus *et al.*, 2004, Mes. Fishes 3, Descr. p. 452. CCIS L2-245, Cretaceous 02, 08

UALVP8707 skull. Forey, 1975, Descr. p. 153. Hermus *et al.*, 2004, Mes. Fishes 3, Descr. p. 452. CCIS L2-245, Cretaceous 02, 08

UALVP8708 anterior part of fish. Forey, 1975, Descr. p. 153. Hermus *et al.*, 2004, Mes. Fishes 3, Descr. p. 452. CCIS L2-245, Cretaceous 02, 11

UALVP8709 skull. Forey, 1975, Descr. p. 153. Hermus *et al.*, 2004, Mes. Fishes 3, Descr. p. 452. CCIS L2-245, Cretaceous 02, 08

UALVP8710 scattered cranial remains. Forey, 1975, Descr. p. 153. Hermus *et al.*, 2004, Mes. Fishes 3, Descr. p. 452. CCIS L2-245, Cretaceous 02, 08

UALVP8711 scattered cranial remains. Forey, 1975, Descr. p. 153. Hermus *et al.*, 2004, Mes. Fishes 3, Descr. p. 452. CCIS L2-245, Cretaceous 02, 08

UALVP8712 skull. Forey, 1975, Descr. p. 153. [**specimen missing as of April 1992**]

UALVP8713 scattered cranial remains a, b. Forey, 1975, Descr. p. 153. Hermus *et al.*, 2004, Mes. Fishes 3, Descr. p. 452. CCIS L2-245, Cretaceous 02, 08

UALVP8714 scattered cranial remains a, b. Forey, 1975, Descr. p. 153. Hermus *et al.*, 2004, Mes. Fishes 3, Descr. p. 452. CCIS L2-245, Cretaceous 02, 08

Osteichthyes Huxley, 1880
   Clupeiformes Bleeker, 1859
     Clupeoidei Bleeker, 1859
       Clupeidae Cuvier, 1817

UALVP48615 Cast of HOLOTYPE Two Casts of Duke University, Primate Center DPC 6202, 3D scans *Chasmoclupea aegyptica* Murray, Simons, and Attia, 2005
Collector: Simons, Elwyn
Africa; Egypt; Fayum Depression: Quarry O, from freshwater deposits.
Tertiary-Oligocene-Oligocene early (Rupelian), Upper Jebel Qatrani Formation. Preserved in three dimensions showing details of the skull roof and lateral skull bones, missing anal and caudal fins, but including pectoral, pelvic, and part of the dorsal fins. Murray *et al.*, 2005, Descr. pp. 301–304 & fig. Fig. 1, p. 302,

Fig. 2, A and B, p. 302, Fig. 3, A and B, p. 303, Fig. 5, A–C, p. 304. CCIS L2-245, Oligocene 01, 04

Murray, Alison M., Simons, Elwyn L., and Attia, Yousry S. 2005. A new clupeid fish (Clupeomorpha) from the Oligocene of Fayum, Egypt, with notes on some other fossil Clupeomorphs. Journal of Vertebrate Paleontology, Vol. 25(2):300–308.

Osteichthyes Huxley, 1880
    Clupeiformes Bleeker, 1859
        Clupeoidei Bleeker, 1859
            Ostariostomidae Schaeffer, 1949
UALVP52610 Cast of HOLOTYPE Cast of Princeton University Geological Museum PU14728. *Ostariostoma wilseyi* **Schaeffer, 1949**
Collector: Wilsey, James A., 1947
The United States: Montana, Madison County, Raw Liver Creek. Cretaceous-Cretaceous Late; or, Tertiary-Paleocene-Danian. Livingston Formation, freshwater varved shales. Cast made from peel of whole fish. Schaeffer, 1949, Descr. pp. 3–15, & fig. Fig. 1A, B. p. 9. CCIS L2-245, Cretaceous 07, 01
    Schaeffer, Bobb 1949. A Teleost from the Livingston Formation of Montana. American Museum Novitates 1427:1–16.

Osteichthyes Huxley, 1880
    Pycnodontiformes Berg, 1940
        Pycnodontidae Agassiz, 1833
UALVP53661 Cast of HOLOTYPE Cast of IGM3286, original at IGM Instituto de Geología, Universidad Nacional Autónoma de México. *Tepexichthys aranguthyorum* **Applegate, 1992, p. 167**
Mexico: Puebla, Tepexi de Rodriguez, Aranguthy portion of the Tlayúa quarry, IGM locality 370. Cretaceous-Cretaceous Early-Albian, Tlayúa Formation. An almost complete fish. Applegate, 1992, Descr. p. 167, & fig. Fig. 3, p. 167, Fig. 4, p. 168, Fig. 8, p. 172. CCIS L2-245, Cretaceous 07, 11

Applegate, Shelton Pleasants 1992. A new genus and species of pycnodont from the Cretaceous (Albian) of central Mexico, Tepexi de Rpodriguez, Puebla. Universidad Nacional Autónoma de México, Instituto de Geología, Revista, Vol. 10(2):164–178.

Osteichthyes Huxley, 1880
    Acanthomorpha (sensu Johnson and Patterson, 1993)
        Aipichthyoidea Otero and Gayet, 1996
        Aipichthyidae Patterson, 1964
UALVP54106 HOLOTYPE *Errachidia pentaspinosa* **Murray and Wilson, 2014, p. 35**
Morocco: Ouarzazate, Agoult. Cretaceous-Cretaceous Late-Turonian, Akrabou Formation. Complete fish in part & counter-part. Murray and Wilson, 2014, Descr. pp. 35–38, Table 1. p. 36, & fig. Fig. 1, A, Fig. 2, p. 37, Fig. 3, A, p. 38.
    Murray, Alison M., and Wilson, Mark V. H. 2014. Four new basal Acanthomorph fishes from the Late Cretaceous of Morocco. Journal of Vertebrate Paleontology, Vol. 34(1):34–48
<u>PARATYPE:</u>
UALVP51611 whole fish. "21." Murray *et al.*, 2013, Descr. p. 540 & fig. Fig. 15, p. 541; Murray and Wilson, 2014, Descr. pp. 35–38, & fig. Fig. 1, B, Fig. 3, B, p. 38, Table 1, p. 36

Osteichthyes Huxley, 1880
    Acanthomorpha (sensu Johnson and Patterson, 1993)
        Aipichthyoidea Otero and Gayet, 1996
        Aipichthyidae Patterson, 1964
UALVP51665 HOLOTYPE *Homalopagus multispinosus* **Murray and Wilson, 2014, p. 38**
Morocco: Ouarzazate, Agoult, site KK0809. Cretaceous-Cretaceous Late-Cenomanian-Turonian, Akrabou Formation. Whole fish. Murray *et al.*, 2013, Descr. p. 540 & fig. Fig. 16, p. 541. Murray and Wilson, 2014, Descr. pp. 38–41, & fig. Fig. 4, A, p. 39, Fig. 5, p. 39. Fig. 6, p. 41, Table 1, p. 36

Murray, Alison M., and Wilson, Mark V. H. 2014. Four new basal Acanthomorph fishes from the Late Cretaceous of Morocco. Journal of Vertebrate Paleontology, Vol. 34(1):34–48
PARATYPE:
UALVP47142 missing the head and tail. Murray *et al.*, 2013, Descr. p. 540–542. Murray and Wilson, 2014, Descr. pp. 38–41, & fig. Fig. 4, B, p. 39, Table 1, p. 36

Osteichthyes Huxley, 1880
    Acanthomorpha (sensu Johnson and Patterson, 1993)
        Polymixiiformes Lowe, 1838
        Pycnosteroididae Patterson, 1964
UALVP51610 HOLOTYPE *Magrebichthys nelsoni* **Murray and Wilson, 2014, p. 41**
Morocco: Ouarzazate, Agoult. Cretaceous-Cretaceous Late-Cenomanian, Akrabou Formation. Whole fish. "19"; Descr. p. 542 & fig. Fig. 17, p. 543. Murray *et al.*, 2013. Murray and Wilson, 2014, Descr. pp. 41–44, & fig. Fig. 7, Fig. 9, Table 2, p. 43
    Murray, Alison M., and Wilson, Mark V. H. 2014. Four new basal Acanthomorph fishes from the Late Cretaceous of Morocco. Journal of Vertebrate Paleontology, Vol. 34(1):34–48
PARATYPES:
UALVP51666 whole fish, part & counterpart. Murray and Wilson, 2014, Descr. pp. 41, 43, 44, Table 2, p. 43; & fig. incorrectly labeled as UALVP51660 in Fig. 8, A, B, p. 42
UALVP51667 whole fish. Murray *et al.*, 2013, Descr. p. 529. Murray and Wilson, 2014, Descr. pp. 41, 43, 44, & fig. Fig. 8, C, Table 2, p. 43
UALVP54108 complete fish. Murray and Wilson, 2014, Descr. pp. 41, 43, 44, & fig. Fig. 8, D, Table 2, p. 43

Osteichthyes Huxley, 1880
    Paracanthopterygii (sensu Patterson and Rosen, 1989)
        Percopsidae Regan, 1911

UALVP13466 HOLOTYPE *Libotonius pearsoni* **Wilson, 1979, p. 401**
Collectors: Lindoe, L. Allan, and Wilson, Mark V. H., 1977
The United States: Washington, Okanogan County, Resner Canyon, 48°54.3'N. Lat., 118°51.2'W. Long., 29 km NNW of Republic. Tertiary-Eocene-Bridgerian NALMA, Klondike Mountain Formation. Complete. Wilson 1979, Descr. p. 401, & fig. Fig. 1A, p. 402. Murray and Wilson, 1999, Descr. p. 400. CCIS L2-245, Eocene 07, 18

Wilson, Mark V. H. 1979. A second species of *Libotonius* (Pisces: Percopsidae) from the Eocene of Washington State. Copeia 1979(3):400–405

PARATYPES:

UALVP13469 complete skeleton. Wilson, 1979, Descr. p. 401, & fig. Fig. 1B, p. 402. Murray and Wilson, 1999, Descr. p. 400. CCIS L2-245, Eocene 07, 18

UALVP14765 part & counterpart. Wilson, 1979, Descr. p. 401. Wilson, 1988, Palaeo., Palaeo., Palaeo. Vol. 62:609–623. Fig. Fig. 8, A, p. 619. Murray and Wilson, 1999, Descr. p. 400. CCIS L2-245, Eocene 07, 18

Osteichthyes Huxley, 1880
    Paracanthopterygii (sensu Patterson and Rosen, 1989)
        Percopsidae Regan, 1911
UALVP30842 a, b, *Massamorichthys wilsoni* **Murray, 1996**
Collector: Lindoe, L. Allan, 1988
Canada: Alberta, Joffre, Joffre Bridge SW Fish Layer
Tertiary-Paleocene, Paskapoo Formation. Part & counterpart. Murray, 1996, Descr. p. 643, premaxilla, dentary, p. 645, (UALVP30842b) hyomandibula, symplectic, p. 646, (UALVP30842b) endopterygoid, p. 647, second centrum, p. 648, (UALVP30842b) hyomandibula pelvic girdle, Fig. 8, p. 649, & Fig. 1A,B, p. 643, meristics, Table 1, p. 645; morphometrics, Table 2, p. 645, scales p. 649. CCIS L2-245, Paleocene 07, 12

Murray, Alison M. 1996. A new Paleocene genus and species of percopsid, *Massamorichthys wilsoni* (Paracanthopterygii) from Joffre Bridge, Alberta, Canada. Journal of Vertebrate Paleontology, Vol. 16(4): 642–652

<u>PARATYPES (* not listed as a paratype but used in original description):</u>

*UALVP21627 almost complete, part & counterpart, no tail. Murray, 1996, Descr. p. 643, parasphenoid, p. 645. CCIS L2-245, Paleocene 07, 11

UALVP21660 skull. Murray, 1996, Descr. p. 643, angular, p. 646, & fig. Fig. 6, p. 647. CCIS L2-245, Paleocene 07, 11

UALVP21758 almost complete tail, part & counterpart, Murray, 1996, Descr. p. 643, infraorbitals. p. 645. CCIS L2-245, Paleocene 07, 11

*UALVP22158 part & counterpart. Murray, 1996, meristics given in Table 1, p. 645, measurements for this specimen given in Table 2, p. 645. CCIS L2-245, Paleocene 07, 11

*UALVP23525 four dorsal fin spines. Murray, 1996, Descr. p. 649

UALVP23534 pectoral fin 4 radials. Murray, 1996, Descr. p. 649. CCIS L2-245, Paleocene 07, 11

UALVP23535 Murray, 1996, Descr. Otolith, p. 644, meristics Table 1, morphometrics Table 2, maxilla, p. 645, first & second centrum, epineurals of first 3 vertebrae, third preural centrum, intermuscular bones, p. 648, fourth preural centrum, dorsal fin spines, pelvic fins, single postcleithrum, p. 649, well-developed parapophyses with ribs on anterior vertebrae, p. 650, & fig. Fig. 3, p. 644, Fig. 7, p. 648. CCIS L2-245, Paleocene 07, 11

UALVP23538 hyomandibula. Murray, 1996, Descr. p. 646. CCIS L2-245, Paleocene 07, 11

UALVP23553 hyomandibula. Murray, 1996, Descr. meristics Table 1, p. 645, p. 646. CCIS L2-245, Paleocene 07, 11

*UALVP25325 Murray, 1996, Descr. meristics, Table 1, p. 645; measurements, Table 2, p. 645, hyomandibula, p. 646, teeth p. 647, scapular foramen, anterior process of coracoid, p. 649. CCIS L2-245, Paleocene 07, 11

*UALVP25417 pelvic fins. Murray, 1996, Descr. p. 649. CCIS L2-245, Paleocene 07, 11

*UALVP25480 subopercle. Murray, 1996, Descr. p. 648. CCIS L2-245, Paleocene 07, 11

*UALVP25614 premaxilla. Murray, 1996, Descr. p. 645

*UALVP25624 Murray, 1996. Descr. measurements, Table 2, p. 645. CCIS L2-245, Paleocene 07, 11

*UALVP25673 Murray, 1996, Descr. meristics, Table 1, measurements, Table 2, p. 645. CCIS L2-245, Paleocene 07, 11

*UALVP25833 Murray, 1996, Descr. meristics, Table 1, p. 645; measurements, Table 2, p. 645, pelvic splint, single postcleithrum, p. 649. CCIS L2-245, Paleocene 07, 11

*UALVP25834 Murray, 1996, Descr. meristics, Table 1, p. 645; measurements, Table 2, p. 645, dorsal fin spines, pterygiophiores, p. 649. CCIS L2-245, Paleocene 07, 11

*UALVP26528 (recatalogued as **UALVP56975**, catalogue number used twice, once for a Mammal, and once for a fish). Murray, 1996, Descr. 2 hypohyals p. 648. CCIS L2-245, Paleocene 07, 11

*UALVP27144 Murray, 1996, Descr. angular, p. 646, palatine p. 647, interopercle, p. 648, dorsal projection running alongside the supracleithrum, single postcleithrum, p. 649. CCIS L2-245, Paleocene 07, 12

*UALVP27145 Murray, 1996, Descr. infraorbitals, retroarticular, p. 645, first centrum, preanal centrum p. 648, pelvic splint, p. 649, well-developed parapophyses with ribs on anterior vertebrae, p. 650. CCIS L2-245, Paleocene 07, 12

*UALVP30846a Murray, 1996, Descr. premaxilla, p. 645, scales, p. 649. CCIS L2-245, Paleocene 07, 12

*UALVP30878a Murray, 1996, Descr. vomer, basioccipital and traces of exoccipitals, premaxilla, p. 645. CCIS L2-245, Paleocene 07, 12

*UALVP30899a Murray, 1996, Descr. supraoccipital p. 644. CCIS L2-245, Paleocene 07, 12

UALVP31683 Murray, 1996, Descr. meristics Table 1, measurements, Table 2, p. 645, third preural centrum, p. 648, dorsal fin spines, dorsal projection running alongside the supracleithrum, p. 649. CCIS L2-245, Paleocene 07, 12

*UALVP32554 Murray, 1996, Descr. pelvic splint, p. 649. CCIS L2-245, Paleocene 07, 12

*UALVP39089 Murray, 1996, Descr. otolith 2.6 mm, head length 13.0 mm p. 644, vomer, parasphenoid, p. 645. CCIS L2-245, Paleocene 07, 12

*UALVP39090 scales. Murray, 1996, Descr. p. 649. CCIS L2-245, Paleocene 07, 12

UALVP39094 Murray, 1996, Descr. skull roof, supraoccipital, frontal, nasals, p. 644, sensory canals, pterotic, lateral ethmoid, vomer, p. 645, & Fig. 4, p. 646, palatine p. 647, dorsal projection running alongside the supracleithrum, p. 649. CCIS L2-245, Paleocene 07, 12

*UALVP56975 (=see *UALVP26528 above) catalogue number used twice, once for a Mammal, and once for a fish. Murray, 1996, Descr. 2 hypohyals p. 648. CCIS L2-245, Paleocene 07, 11

Osteichthyes Huxley, 1880
    Paracanthopterygii (sensu Patterson and Rosen, 1989)
      Percopsidae Regan, 1911
UALVP34771 HOLOTYPE *Lateopisciculus turrifumosus* **Murray and Wilson, 1996, p. 429**
Collector: Tieman, Bert, 1978
Canada: Alberta, Smoky Tower, Smoky Tower #1. Tertiary-Paleocene, Paskapoo Formation. Complete fish skeleton. Murray and Wilson, 1996, fig. Fig. 1, p. 431, Tables 1, 2, p. 431. Murray and Wilson, 1999, Mesozoic Fishes 2- Systematics and Fossil Record, p. 400 as *Lateopisciculus turrifumosis*. Newbrey *et al.*, 2013, Descr. pp. 364, 370. CCIS L2-245, Paleocene 06, 07

Murray, A. M., and Wilson, M. V. H. 1996. A new Palaeocene genus and species of percopsiform (Teleostei: Paracanthopterygii) from the Paskapoo Formation, Smoky Tower, Alberta. Canadian Journal of Earth Sciences, Vol. 33:429–438

PARATYPES:

UALVP21541.1 postcranial fish skeleton. Murray and Wilson, 1996, Tables 1, 2, p. 431. Murray and Wilson, 1999, Mesozoic Fishes 2-Systematics and Fossil Record, p. 400 as *Lateopisciculus turrifumosis* [Note: UALVP21541 used for two separate paratypes, *Lateopisciculus turrifumosus*. Murray and Wilson, 1996; and, *Speirsaenigma lindoei* Wilson and Williams, 1991, note different localities]. CCIS L2-245, Paleocene 06, 07

UALVP22870 complete, part & counterpart; found in same slab as *Esox tiemani* types. Murray and Wilson, 1996, fig. Fig. 2, p. 432, Fig. 3, p. 433, Fig. 4a, p. 434, Fig. 6, p. 436, Tables 1, 2, p. 431; published as *Lateopisciculus turrifumosis* in Murray and Wilson, 1999, Mesozoic Fishes 2- Systematics and Fossil Record, p. 400. Newbrey *et al.*, 2013, Descr. p. 364. CCIS L2-245, Paleocene 06, 07

UALVP34772 complete fish skeleton. Murray and Wilson, 1996, published as *Lateopisciculus turrifumosus*, Tables 1, 2 p. 431. Murray and Wilson, 1999, Mesozoic Fishes 2-Systematics and Fossil Record, published as *Lateopisciculus turrifumosis*, p. 400. CCIS L2-245, Paleocene 06, 07

Osteichthyes Huxley, 1880

    Paracanthopterygii (sensu Patterson and Rosen, 1989)

        Sphenocephaliformes Rosen and Patterson, 1969 (as Sphenocephaloidei)

            Sphenocephalidae Patterson, 1964

UALVP32133 HOLOTYPE *Xenyllion zonensis* **Wilson and Murray, 1996, p. 372**

Collector: L. Allan Lindoe, 1989.

Canada: Alberta: Smoky River. Cretaceous, Upper Albian, Shaftesbury Formation, shale above Fish Scale Zone Disarticulated but associated skull preserved in part & counterpart. Wilson, and

Murray, 1996, Descr. pp. 372, 374, 376, 378, & 379, & fig. Fig. 2, a, b, c, d, p 373; Fig. 3, a-i, p. 374. Newbrey *et al.* 2013, Descr. pp. 364, 368, 370, 372, 375, Table 1, p. 377, & fig. Fig. 4, G, p. 370–371, Fig. 8, A, p. 378. CCIS L2-245 Cretaceous 6, 14

Wilson, Mark V. H., and Murray, Alison M. 1996. Early Cenomanian acanthomorph teleost in the Cretaceou Fish Scale Zone, Albian/Cenomanian boundary, Alberta, Canada. pp. 369–382. *IN:* Arratia, G., & Viohl, G. (editors), Mesozoic Fishes – Systematics and Paleoecology. Dr. Friedrich Pfeil. München, Germany. 576 pp.

<u>PARATYPES:</u>

UALVP32073 left opercle. Wilson and Murray, 1996, Descr. p. 372. Newbrey *et al.*, 2013, Descr. pp. 364, 370. CCIS L2-245 Cretaceous 6, 15

UALVP32086 dentary. Wilson and Murray, 1996, Descr. p. 372. Newbrey *et al.*, 2013, Descr. pp. 364, 370. CCIS L2-245 Cretaceous 6, 15

UALVP32095 preopercle. Wilson and Murray, 1996, Descr. pp. 372, 376, & fig. Fig. 4, e, p. 377. Newbrey *et al.*, 2013, Descr. pp. 364, 372, 374. CCIS L2-245 Cretaceous 6, 15

UALVP32127 premaxilla with teeth. Wilson and Murray, 1996, Descr. pp. 372, 374, & fig. Fig. 4, a, p. 377. CCIS L2-245 Cretaceous 6, 14

UALVP32131 right opercle. Wilson and Murray, 1996, Descr. pp. 372, 378. Newbrey *et al.*, 2013, Descr. pp. 364 (said right opercle), p. 374. & fig. Fig. 4, E, p. 370 (said left opercle). CCIS L2-245 Cretaceous 3, 12

UALVP32182 preopercle. Wilson and Murray, 1996, Descr. p. 372. CCIS L2-245 Cretaceous 3, 12

UALVP32185 dentary. Wilson and Murray, 1996, Descr. pp. 372, 374, & fig. Fig. 4, b, c, p. 377. Newbrey *et al.*, 2013, Descr. p. 364, 370. CCIS L2-245 Cretaceous 3, 12

UALVP32197 opercle. Wilson and Murray, 1996, Descr. pp. 372, 378, & fig. Fig. 4, g, p. 377. Newbrey *et al.*, 2013, Descr. p. 370. CCIS L2-245 Cretaceous 6, 13

UALVP32227 right preopercle. Wilson and Murray, 1996, Descr. on p. 372 as a preopercle, Descr. on p. 374 as an "isolated frontal." Newbrey *et al.*, 2013, Descr. p. 364. CCIS L2-245 Cretaceous 6, 15

UALVP32228 frontal. Wilson and Murray, 1996, Descr. p. 372. CCIS L2-245 Cretaceous 6, 15

UALVP32231 possible posttemporal. Wilson and Murray, 1996, Descr. pp. 372, & fig. Fig. 4, f, p. 377. CCIS L2-245 Cretaceous 3, 12

UALVP32256 frontal. Wilson and Murray, 1996, Descr. pp. 372, 374, & fig. Fig. 4, d, p. 377. CCIS L2-245 Cretaceous 3, 13 & Cretaceous 3, 14

UALVP39071 angulo-articular. Wilson and Murray, 1996, Descr. on p. 372. CCIS L2-245 Cretaceous 6, 15

Osteichthyes Huxley, 1880
    Ostariophysi Sagemehl, 1885
        Cypriniformes Bleeker, 1859
            Catostomidae Bonaparte, 1840
UALVP13407 Cast of HOLOTYPE *Amyzon mentale* **Cope, 1872**
Collector: Camp, R. R.
The United States: Nevada, Elko. Tertiary-Oligocene, *Amyzon* Shale. CCIS L2-245, Oligocene 01, 02
    Cope, Edward Drinker 1872. On the Tertiary coal and fossils of Osino, Nevada. Proceedings, American Philosophical Society, Vol. 12:478–481

Osteichthyes Huxley, 1880
    Ostariophysi Sagemehl, 1885
        Cypriniformes Bleeker, 1859
            Catostomidae Bonaparte, 1840
UALVP55260 HOLOTYPE *Amyzon kishenehnicum* **Liu, Wilson, and Murray, 2016, p. 290**
The United States: Montana, Disbrow Creek. Eocene, Kishenehn Formation. Complete skeleton. Liu *et al.*, 2016, Fig. 1, p. 291. CCIS L2-245, Eocene 21, 04

Liu, Juan, Wilson, Mark V. H., and Murray, Alison M. 2016. A new catostomid fish (Ostariophysi, Cypriniformes) from the Eocene Kishenehn Formation and remarks on the North American species of †*Amyzon* Cope, 1872. Journal of Paleontology, Vol. 90(2):288–304

PARATYPES:

UALVP23943 fish (on back of UALVP24141-3). Liu *et al.*, 2016, fig. Fig. 3, p. 293

UALVP24131 fish. Liu *et al.*, 2016, fig. Fig. 3, p. 293. CCIS L2-245, Eocene 18, 16

UALVP24132 fish. Liu *et al.*, 2016, fig. Fig. 3, p. 293. CCIS L2-245, Eocene 18, 16

UALVP24133 fish. Liu *et al.*, 2016, fig. Fig. 3, p. 293. CCIS L2-245, Eocene 18, 16

UALVP24134 fish. Liu *et al.*, 2016, fig. Fig. 3, p. 293. CCIS L2-245, Eocene 18, 16

UALVP24137 complete fish in left lateral view. Liu *et al.*, 2016, Descr. p. 291, & fig. Fig. 2.1, p. 292, Fig. 3, p. 293. CCIS L2-245, Eocene 18, 17

UALVP24138 fish. Liu *et al.*, 2016, fig. Fig. 3, p. 293. CCIS L2-245, Eocene 18, 17

UALVP24139 fish. Liu *et al.*, 2016, fig. Fig. 3, p. 293. CCIS L2-245, Eocene 18, 17

UALVP24140 nearly complete juvenile fish. Liu *et al.*, 2016, Descr. p. 291, & fig. Fig. 3, p. 293. CCIS L2-245, Eocene 18, 17

UALVP24141 fish. Liu *et al.*, 2016, fig. Fig. 3, p. 293. CCIS L2-245, Eocene 18, 17

UALVP24142 fish. Liu *et al.*, 2016, fig. Fig. 3, p. 293. CCIS L2-245, Eocene 18, 17

UALVP24143 fish. Liu *et al.*, 2016, fig. Fig. 3, p. 293. CCIS L2-245, Eocene 18, 17

UALVP24144 fish. Liu *et al.*, 2016, fig. Fig. 3, p. 293. CCIS L2-245, Eocene 18, 17

UALVP24145 fish. Liu *et al.*, 2016, fig. Fig. 3, p. 293. CCIS L2-245, Eocene 18, 17

UALVP24146 fish. Liu *et al.* 2016 fig. Fig. 3, p. 293. CCIS L2-245, Eocene 18, 17

UALVP24147 nearly complete juvenile fish. Liu *et al.*, 2016, Descr. p. 291, & fig. Fig. 3, p. 293. CCIS L2-245, Eocene 18, 17

UALVP24148 nearly complete juvenile fish. Liu *et al.*, 2016, Descr. p. 291, & fig. Fig. 3, p. 293. CCIS L2-245, Eocene 18, 17

UALVP24149 nearly complete juvenile fish. Liu *et al.*, 2016, Descr. p. 291, & fig. Fig. 3, p. 293. CCIS L2-245, Eocene 18, 17

UALVP24150 fish. Liu *et al.*, 2016, fig. Fig. 3, p. 293. CCIS L2-245, Eocene 18, 17

UALVP24151 fish. Liu *et al.*, 2016, fig. Fig. 3, p. 293. CCIS L2-245, Eocene 18, 17

UALVP24152 nearly complete juvenile fish. Liu *et al.*, 2016, Descr. p. 291, & fig. Fig. 3, p. 293. CCIS L2-245, Eocene 18, 17

UALVP24153 fish. Liu *et al.* 2016 fig. Fig. 3, p. 293. CCIS L2-245, Eocene 18, 17

UALVP24154 nearly complete juvenile fish. Liu *et al.*, 2016, Descr. p. 291, & fig. Fig. 3, p. 293. CCIS L2-245, Eocene 18, 17

UALVP24155 fish. Liu *et al.*, 2016, fig. Fig. 3, p. 293. CCIS L2-245, Eocene 18, 17

UALVP24156 fish. Liu *et al.*, 2016, fig. Fig. 3, p. 293. CCIS L2-245, Eocene 18, 17

UALVP24157 fish. Liu *et al.*, 2016, fig. Fig. 3, p. 293. CCIS L2-245, Eocene 18, 17

UALVP24158 fish. Liu *et al.*, 2016, fig. Fig. 3, p. 293. CCIS L2-245, Eocene 21, 4

UALVP24159 fish. Liu *et al.*, 2016, fig. Fig. 3, p. 293. CCIS L2-245, Eocene 18, 17

UALVP24160 fish. Liu *et al.*, 2016, fig. Fig. 3, p. 293. CCIS L2-245, Eocene 18, 16

UALVP24161 fish. Liu *et al.*, 2016, fig. Fig. 3, p. 293. CCIS L2-245, Eocene 18, 16

UALVP24162 fish. Liu *et al.*, 2016, fig. Fig. 3, p. 293. CCIS L2-245, Eocene 18, 16

UALVP24163 fish. Liu *et al.*, 2016, fig. Fig. 3, p. 293. CCIS L2-245, Eocene 18, 16

UALVP24164 fish. Liu *et al.*, 2016, fig. Fig. 3, p. 293. CCIS L2-245, Eocene 18, 16

UALVP24165 fish. Liu *et al.*, 2016, fig. Fig. 3, p. 293. CCIS L2-245, Eocene 18, 16

UALVP24166 fish. Liu *et al.*, 2016, fig. Fig. 3, p. 293. CCIS L2-245, Eocene 18, 16

UALVP24167 fish. Liu *et al.*, 2016, fig. Fig. 3, p. 293. CCIS L2-245, Eocene 18, 16

UALVP24168 fish. Liu *et al.*, 2016, fig. Fig. 3, p. 293. CCIS L2-245, Eocene 18, 16

UALVP24169 fish. Liu *et al.*, 2016, fig. Fig. 3, p. 293. CCIS L2-245, Eocene 18, 16

UALVP24170 fish. Liu *et al.*, 2016, fig. Fig. 3, p. 293. CCIS L2-245, Eocene 18, 16

UALVP24171 fish. Liu *et al.*, 2016, fig. Fig. 3, p. 293. CCIS L2-245, Eocene 18, 16

UALVP24172 fish. Liu *et al.*, 2016, fig. Fig. 3, p. 293. CCIS L2-245, Eocene 18, 16

UALVP24173 fish. Liu *et al.*, 2016, fig. Fig. 3, p. 293. CCIS L2-245, Eocene 18, 16

UALVP24174 fish. Liu *et al.*, 2016, fig. Fig. 3, p. 293. CCIS L2-245, Eocene 18, 16

UALVP24175 fish. Liu *et al.*, 2016, fig. Fig. 3, p. 293. CCIS L2-245, Eocene 18, 16

UALVP24176 fish. Liu *et al.*, 2016, fig. Fig. 3, p. 293. CCIS L2-245, Eocene 18, 16

UALVP24177 fish. Liu *et al.*, 2016, fig. Fig. 3, p. 293. CCIS L2-245, Eocene 18, 16

UALVP24178 fish. Liu *et al.*, 2016, fig. Fig. 3, p. 293. CCIS L2-245, Eocene 18, 16

UALVP24179 fish. Liu *et al.*, 2016, fig. Fig. 3, p. 293. CCIS L2-245, Eocene 18, 16

UALVP24180 fish. Liu *et al.*, 2016, fig. Fig. 3, p. 293. CCIS L2-245, Eocene 18, 16

UALVP24181 fish. Liu *et al.*, 2016, fig. Fig. 3, p. 293. CCIS L2-245, Eocene 18, 16

UALVP24182 fish. Liu *et al.*, 2016, fig. Fig. 3, p. 293. CCIS L2-245, Eocene 18, 16

UALVP24183 fish. Liu *et al.*, 2016, fig. Fig. 3, p. 293. CCIS L2-245, Eocene 18, 16

UALVP24184 fish. Liu *et al.*, 2016, fig. Fig. 3, p. 293. CCIS L2-245, Eocene 18, 16

UALVP24185 fish. Liu *et al.*, 2016, fig. Fig. 3, p. 293. CCIS L2-245, Eocene 18, 16

UALVP24186 fish. Liu *et al.*, 2016, fig. Fig. 3, p. 293. CCIS L2-245, Eocene 18, 16

UALVP24187 fish. Liu *et al.*, 2016, fig. Fig. 3, p. 293. CCIS L2-245, Eocene 18, 16

UALVP24188 fish. Liu *et al.*, 2016, fig. Fig. 3, p. 293. CCIS L2-245, Eocene 18, 16

UALVP24189 fish. Liu *et al.*, 2016, fig. Fig. 3, p. 293. CCIS L2-245, Eocene 18, 16

UALVP24190 fish. Liu *et al.*, 2016, fig. Fig. 3, p. 293. CCIS L2-245, Eocene 18, 16

UALVP24191 fish. Liu *et al.*, 2016, fig. Fig. 3, p. 293. CCIS L2-245, Eocene 18, 16

UALVP24192 fish. Liu *et al.*, 2016, fig. Fig. 3, p. 293. CCIS L2-245, Eocene 18, 16

UALVP24193 fish. Liu *et al.*, 2016, fig. Fig. 3, p. 293. CCIS L2-245, Eocene 21, 4

UALVP24194 fish. Liu *et al.*, 2016, fig. Fig. 3, p. 293. CCIS L2-245, Eocene 18, 16

UALVP24195 fish. Liu *et al.*, 2016, fig. Fig. 3, p. 293. CCIS L2-245, Eocene 21, 4

UALVP24196 fish. Liu *et al.*, 2016, fig. Fig. 3, p. 293. CCIS L2-245, Eocene 18, 16

UALVP24197 fish. Liu *et al.*, 2016, fig. Fig. 3, p. 293. CCIS L2-245, Eocene 18, 16

UALVP24198 fish. Liu *et al.*, 2016, fig. Fig. 3, p. 293. CCIS L2-245, Eocene 18, 16

UALVP24199 fish. Liu *et al.*, 2016, fig. Fig. 3, p. 293. CCIS L2-245, Eocene 18, 16

UALVP24226 fish. Liu *et al.*, 2016, fig. Fig. 3, p. 293. CCIS L2-245, Eocene 21, 4

UALVP38728 fish. Liu *et al.*, 2016, fig. Fig. 3, p. 293. CCIS L2-245, Eocene 18, 19

UALVP38729 fish (on slab with UALVP38728). Liu *et al.*, 2016, fig. Fig. 3, p. 293. CCIS L2-245, Eocene 18, 19

UALVP38730 fish. Liu *et al.*, 2016, fig. Fig. 3, p. 293. CCIS L2-245, Eocene 18, 19

UALVP38731 fish. Liu *et al.*, 2016, fig. Fig. 3, p. 293. CCIS L2-245, Eocene 18, 19

UALVP38732 fish. Liu *et al.*, 2016, fig. Fig. 3, p. 293. CCIS L2-245, Eocene 18, 19

UALVP38733 fish part & counterpart. Liu *et al.*, 2016, fig. Fig. 3, p. 293. CCIS L2-245, Eocene 18, 19

UALVP38740 fish part & counterpart. Liu *et al.*, 2016, fig. Fig. 3, p. 293. CCIS L2-245, Eocene 21, 4

UALVP38784 fish. Liu *et al.*, 2016, fig. Fig. 3, p. 293. CCIS L2-245, Eocene 20, 03

UALVP38785 fish. Liu *et al.*, 2016, fig. Fig. 3, p. 293. CCIS L2-245, Eocene 20, 03

UALVP38786 fish. Liu *et al.*, 2016, fig. Fig. 3, p. 293. CCIS L2-245, Eocene 20, 03

UALVP38787 fish. Liu *et al.*, 2016, fig. Fig. 3, p. 293. CCIS L2-245, Eocene 20, 03

UALVP38788 fish. Liu *et al.*, 2016, fig. Fig. 3, p. 293. CCIS L2-245, Eocene 18, 14

UALVP38792 fish. Liu *et al.*, 2016, fig. Fig. 3, p. 293. CCIS L2-245, Eocene 21, 03

UALVP38793 fish. Liu *et al.*, 2016, fig. Fig. 3, p. 293. CCIS L2-245, Eocene 21, 03

UALVP38794 fish. Liu *et al.*, 2016, fig. Fig. 3, p. 293. CCIS L2-245, Eocene 21, 03

UALVP38795 fish. Liu *et al.*, 2016, fig. Fig. 3, p. 293. CCIS L2-245, Eocene 21, 03

UALVP38796 fish. Liu *et al.*, 2016, fig. Fig. 3, p. 293. CCIS L2-245, Eocene 21, 03

UALVP38797 part & counterpart fish. Liu *et al.*, 2016, fig. Fig. 3, p. 293. CCIS L2-245, Eocene 21, 03

UALVP38798 part & counterpart fish. Liu *et al.*, 2016, fig. Fig. 3, p. 293. CCIS L2-245, Eocene 20, 01

UALVP38799 part & counterpart fish. Liu *et al.*, 2016, fig. Fig. 3, p. 293. CCIS L2-245, Eocene 20, 01

UALVP38800 fish. Liu *et al.*, 2016, fig. Fig. 3, p. 293. CCIS L2-245, Eocene 20, 01

UALVP38801 fish. Liu *et al.*, 2016, fig. Fig. 3, p. 293. CCIS L2-245, Eocene 20, 01

UALVP38802 fish. Liu *et al.*, 2016, fig. Fig. 3, p. 293. CCIS L2-245, Eocene 20, 01

UALVP38803 part & counterpart fish. Liu *et al.*, 2016, fig. Fig. 3, p. 293. CCIS L2-245, Eocene 20, 01

UALVP38804 fish. Liu *et al.*, 2016, fig. Fig. 3, p. 293. CCIS L2-245, Eocene 20, 01

UALVP38805 fish. Liu *et al.*, 2016, fig. Fig. 3, p. 293. CCIS L2-245, Eocene 20, 01

UALVP38806 fish. Liu *et al.*, 2016, fig. Fig. 3, p. 293. CCIS L2-245, Eocene 20, 01

UALVP38807 counterpart is in paleo insect collection on backside of a wasp, fossil insect coll. UAPC7284; fish. Liu *et al.*, 2016, fig. Fig. 3, p. 293. CCIS L2-245, Eocene 20, 01

UALVP38876 fish. Liu *et al.*, 2016, fig. Fig. 3, p. 293

UALVP38877 fish. Liu *et al.*, 2016, fig. Fig. 3, p. 293. CCIS L2-245, Eocene 18, 07

UALVP38881 fish. Liu *et al.*, 2016, fig. Fig. 3, p. 293. CCIS L2-245, Eocene 18, 07

UALVP38882 part & counterpart fish. Liu *et al.*, 2016, fig. Fig. 3, p. 293. CCIS L2-245, Eocene 18, 07

UALVP38883 part & counterpart fish. Liu *et al.*, 2016, fig. Fig. 3, p. 293. CCIS L2-245, Eocene 18, 07

UALVP38884 part & counterpart; fragmented. Liu *et al.*, 2016, fig. Fig. 3, p. 293. CCIS L2-245, Eocene 18, 10

UALVP38885 part & counterpart. Liu *et al.*, 2016, fig. Fig. 3, p. 293. CCIS L2-245, Eocene 18, 10

UALVP38886 part & counterpart. Liu *et al.*, 2016, fig. Fig. 3, p. 293. CCIS L2-245, Eocene 18, 10

UALVP38887 part & counterpart. Liu *et al.*, 2016, fig. Fig. 3, p. 293. CCIS L2-245, Eocene 18, 09

UALVP38888 part & counterpart. Liu *et al.*, 2016, fig. Fig. 3, p. 293. CCIS L2-245, Eocene 18, 09

UALVP38889 part & counterpart. Liu *et al.*, 2016, fig. Fig. 3, p. 293. CCIS L2-245, Eocene 18, 09

UALVP38890 Liu *et al.*, 2016, fig. Fig. 3, p. 293. CCIS L2-245, Eocene 18, 11

UALVP38898 Liu *et al.*, 2016, fig. Fig. 3, p. 293. CCIS L2-245, Eocene 18, 04

UALVP38900 Liu *et al.*, 2016, fig. Fig. 3, p. 293. CCIS L2-245, Eocene 18, 02

UALVP38901 Liu *et al.*, 2016, fig. Fig. 3, p. 293. CCIS L2-245, Eocene 18, 02

UALVP38902 Liu *et al.*, 2016, fig. Fig. 3, p. 293. CCIS L2-245, Eocene 18, 02

UALVP38903 Liu *et al.*, 2016, fig. Fig. 3, p. 293. CCIS L2-245, Eocene 18, 02

UALVP38904 Liu *et al.*, 2016, fig. Fig. 3, p. 293. CCIS L2-245, Eocene 18, 02

UALVP38905 Liu *et al.*, 2016, fig. Fig. 3, p. 293. CCIS L2-245, Eocene 18, 02

UALVP38907 Liu *et al.*, 2016, fig. Fig. 3, p. 293. CCIS L2-245, Eocene 18, 02

UALVP38908 Liu *et al.*, 2016, fig. Fig. 3, p. 293. CCIS L2-245, Eocene 18, 04

UALVP38909 Liu *et al.*, 2016, fig. Fig. 3, p. 293. CCIS L2-245, Eocene 18, 02

UALVP38910 Liu *et al.*, 2016, fig. Fig. 3, p. 293. CCIS L2-245, Eocene 18, 02

UALVP38911 Liu *et al.*, 2016, fig. Fig. 3, p. 293. CCIS L2-245, Eocene 18, 02

UALVP38912 Liu *et al.*, 2016, fig. Fig. 3, p. 293. CCIS L2-245, Eocene 18, 02

UALVP38913 Liu *et al.*, 2016, fig. Fig. 3, p. 293. CCIS L2-245, Eocene 18, 02

UALVP38914 Liu *et al.*, 2016, fig. Fig. 3, p. 293. CCIS L2-245, Eocene 18, 02

UALVP38915 head, anterior trunk. Liu *et al.*, 2016 fig. Fig. 3, p. 293. CCIS L2-245, Eocene 18, 11

UALVP38916 scale. Liu *et al.*, 2016, fig. Fig. 3, p. 293. CCIS L2-245, Eocene 18, 11

UALVP38917 scale. Liu *et al.*, 2016, fig. Fig. 3, p. 293. CCIS L2-245, Eocene 18, 11

UALVP38918 Liu *et al.*, 2016, fig. Fig. 3, p. 293. CCIS L2-245, Eocene 18, 11

UALVP38919 part & counterpart; tail, posterior trunk. Liu *et al.*, 2016, fig. Fig. 3, p. 293. CCIS L2-245, Eocene 18, 11

UALVP38920 Liu *et al.*, 2016, fig. Fig. 3, p. 293. CCIS L2-245, Eocene 18, 11

UALVP38922 Liu *et al.*, 2016, fig. Fig. 3, p. 293. CCIS L2-245, Eocene 18, 05

UALVP38923 Liu *et al.*, 2016, fig. Fig. 3, p. 293

UALVP38924 part & counterpart; tail, posterior trunk. Liu *et al.*, 2016, fig. Fig. 3, p. 293. CCIS L2-245, Eocene 18, 05

UALVP38925 Liu *et al.*, 2016, fig. Fig. 3, p. 293. CCIS L2-245, Eocene 18, 03

UALVP38926 Liu *et al.*, 2016, fig. Fig. 3, p. 293. CCIS L2-245, Eocene 18, 05

UALVP38927 Liu *et al.*, 2016, fig. Fig. 3, p. 293. CCIS L2-245, Eocene 18, 05

UALVP38928 Liu *et al.*, 2016, fig. Fig. 3, p. 293. CCIS L2-245, Eocene 18, 05

UALVP38929 on slab with UALVP38930. Liu *et al.*, 2016, fig. Fig. 3, p. 293. CCIS L2-245, Eocene 18, 05

UALVP38930 on slab with UALVP38929. Liu *et al.*, 2016, fig. Fig. 3, p. 293. CCIS L2-245, Eocene 18, 05

UALVP38931 Liu *et al.*, 2016, fig. Fig. 3, p. 293. CCIS L2-245, Eocene 18, 05

UALVP38932 Liu *et al.*, 2016, fig. Fig. 3, p. 293. CCIS L2-245, Eocene 18, 05

UALVP38933 part & counterpart. Liu *et al.*, 2016, fig. Fig. 3, p. 293. CCIS L2-245, Eocene 18, 05

UALVP38934 Liu *et al.*, 2016, fig. Fig. 3, p. 293. CCIS L2-245, Eocene 18, 05

UALVP38935 Liu *et al.*, 2016, fig. Fig. 3, p. 293. CCIS L2-245, Eocene 18, 05

UALVP38936 Liu *et al.*, 2016, fig. Fig. 3, p. 293. CCIS L2-245, Eocene 18, 13

UALVP38937 Liu *et al.*, 2016, fig. Fig. 3, p. 293. CCIS L2-245, Eocene 18, 13

UALVP38938 Liu *et al.*, 2016, fig. Fig. 3, p. 293. CCIS L2-245, Eocene 18, 13

UALVP38939 Liu *et al.*, 2016, fig. Fig. 3, p. 293. CCIS L2-245, Eocene 18, 13

UALVP38940 Liu *et al.*, 2016, fig. Fig. 3, p. 293. CCIS L2-245, Eocene 18, 13

UALVP38941 Liu *et al.*, 2016, fig. Fig. 3, p. 293. CCIS L2-245, Eocene 18, 13

UALVP38942 Liu *et al.*, 2016, fig. Fig. 3, p. 293. CCIS L2-245, Eocene 18, 13

UALVP38943 Liu *et al.*, 2016, fig. Fig. 3, p. 293. CCIS L2-245, Eocene 18, 13

UALVP38944 Liu *et al.*, 2016, fig. Fig. 3, p. 293. CCIS L2-245, Eocene 18, 13

UALVP38945 partial trunk. Liu *et al.*, 2016, fig. Fig. 3, p. 293. CCIS L2-245, Eocene 18, 13

UALVP38946 Liu *et al.*, 2016, fig. Fig. 3, p. 293. CCIS L2-245, Eocene 18, 13

UALVP38947 Liu *et al.*, 2016, fig. Fig. 3, p. 293. CCIS L2-245, Eocene 18, 13

UALVP38948 Liu *et al.*, 2016, fig. Fig. 3, p. 293. CCIS L2-245, Eocene 18, 13

UALVP38949 Liu *et al.*, 2016, fig. Fig. 3, p. 293. CCIS L2-245, Eocene 18, 13

UALVP38950 Liu *et al.*, 2016, fig. Fig. 3, p. 293. CCIS L2-245, Eocene 18, 13

UALVP38951 tail missing. Liu *et al.*, 2016, fig. Fig. 3, p. 293. CCIS L2-245, Eocene 18, 13

UALVP38952 Liu *et al.*, 2016, fig. Fig. 3, p. 293. CCIS L2-245, Eocene 18, 13

UALVP38953 Liu *et al.*, 2016, fig. Fig. 3, p. 293. CCIS L2-245, Eocene 18, 13

UALVP38954 Liu *et al.*, 2016, fig. Fig. 3, p. 293. CCIS L2-245, Eocene 18, 13

UALVP38955 tail missing. Liu *et al.*, 2016, fig. Fig. 3, p. 293. CCIS L2-245, Eocene 18, 13

UALVP38956 head missing. Liu *et al.*, 2016, fig. Fig. 3, p. 293

UALVP38957 Liu *et al.*, 2016, fig. Fig. 3, p. 293. CCIS L2-245, Eocene 18, 13

UALVP38958 part & counterpart; head, anterior trunk. Liu *et al.*, 2016, fig. Fig. 3, p. 293. CCIS L2-245, Eocene 18, 13

UALVP38962 part & counterpart. Liu *et al.*, 2016, fig. Fig. 3. CCIS L2-245, Eocene 18, 01

UALVP38963 bones. Liu *et al.*, 2016, fig. Fig. 3, p. 293. CCIS L2-245, Eocene 18, 01

UALVP38964 Liu *et al.*, 2016, fig. Fig. 3. CCIS L2-245, Eocene 18, 01

UALVP38965 part & counterpart tail, anterior trunk. Liu *et al.*, 2016, fig. Fig. 3, p. 293. CCIS L2-245, Eocene 18, 01

UALVP38966 part & counterpart incomplete. Liu *et al.*, 2016, fig. Fig. 3, p. 293. CCIS L2-245, Eocene 18, 01

UALVP38967 Liu *et al.*, 2016, fig. Fig. 3, p. 293. CCIS L2-245, Eocene 18, 15

UALVP38968 Liu *et al.*, 2016, fig. Fig. 3, p. 293. CCIS L2-245, Eocene 18, 15

UALVP38969 Liu *et al.*, 2016, fig. Fig. 3, p. 293. CCIS L2-245, Eocene 18, 15

UALVP38970 part & partial counterpart. Liu *et al.*, 2016, fig. Fig. 3, p. 293. CCIS L2-245, Eocene 18, 15

UALVP38971 Liu *et al.*, 2016, fig. Fig. 3, p. 293. CCIS L2-245, Eocene 18, 15

UALVP38972 Liu *et al.*, 2016, fig. Fig. 3, p. 293. CCIS L2-245, Eocene 18, 15

UALVP38973 Liu *et al.*, 2016, fig. Fig. 3, p. 293. CCIS L2-245, Eocene 18, 15

UALVP38974 on slab with UALVP38975. Liu *et al.*, 2016, fig. Fig. 3, p. 293. CCIS L2-245, Eocene 18, 15

UALVP38975 on slab with UALVP38974. Liu *et al.*, 2016, fig. Fig. 3, p. 293. CCIS L2-245, Eocene 18, 15

UALVP38976 Liu *et al.*, 2016, fig. Fig. 3, p. 293. CCIS L2-245, Eocene 18, 14

UALVP38977 partial. Liu *et al.*, 2016, fig. Fig. 3, p. 293. CCIS L2-245, Eocene 18, 14

UALVP38978 partial; on slab with UALVP38979. Liu *et al.*, 2016, fig. Fig. 3, p. 293. CCIS L2-245, Eocene 18, 14

UALVP38979 partial; on slab with UALVP38978. Liu *et al.*, 2016, fig. Fig. 3, p. 293. CCIS L2-245, Eocene 18, 14

UALVP38980 Liu *et al.*, 2016, fig. Fig. 3, p. 293. CCIS L2-245, Eocene 18, 14

UALVP38981 no tail. Liu *et al.*, 2016, fig. Fig. 3, p. 293. CCIS L2-245, Eocene 18, 14

UALVP38982 part & counterpart. Liu *et al.*, 2016, fig. Fig. 3, p. 293. CCIS L2-245, Eocene 18, 14

UALVP38983 Liu *et al.*, 2016, fig. Fig. 3, p. 293. CCIS L2-245, Eocene 18, 14

UALVP38984 Liu *et al.*, 2016, fig. Fig. 3, p. 293. CCIS L2-245, Eocene 18, 14

UALVP38985 Liu *et al.*, 2016, fig. Fig. 3, p. 293. CCIS L2-245, Eocene 18, 14

UALVP38986 head missing. Liu *et al.*, 2016, fig. Fig. 3, p. 293. CCIS L2-245, Eocene 18, 14

UALVP38987 head missing. Liu *et al.*, 2016, fig. Fig. 3, p. 293. CCIS L2-245, Eocene 18, 14

UALVP38988 Liu *et al.*, 2016, fig. Fig. 3, p. 293

UALVP38989 Liu *et al.*, 2016, fig. Fig. 3, p. 293

UALVP38990 part & counterpart. Liu *et al.*, 2016, fig. Fig. 3, p. 293. CCIS L2-245, Eocene 18, 14

UALVP38991 head missing. Liu *et al.*, 2016, fig. Fig. 3, p. 293. CCIS L2-245, Eocene 18, 14

UALVP38992 Liu *et al.*, 2016, fig. Fig. 3, p. 293. CCIS L2-245, Eocene 18, 14

UALVP38993 Liu *et al.*, 2016, fig. Fig. 3, p. 293. CCIS L2-245, Eocene 18, 14

UALVP38994 Liu *et al.*, 2016, fig. Fig. 3, p. 293. CCIS L2-245, Eocene 18, 14

UALVP38995 Liu *et al.*, 2016, fig. Fig. 3, p. 293. CCIS L2-245, Eocene 18, 06

UALVP38996 Liu *et al.*, 2016, fig. Fig. 3, p. 293. CCIS L2-245, Eocene 18, 06

UALVP38997 Liu *et al.*, 2016, fig. Fig. 3, p. 293. CCIS L2-245, Eocene 18, 06

UALVP38998 Liu *et al.*, 2016, fig. Fig. 3, p. 293. CCIS L2-245, Eocene 18, 06

UALVP38999 Liu *et al.*, 2016, fig. Fig. 3, p. 293. CCIS L2-245, Eocene 18, 06

UALVP39000 Liu *et al.*, 2016, fig. Fig. 3, p. 293. CCIS L2-245, Eocene 18, 02

UALVP39001 Liu *et al.*, 2016, fig. Fig. 3, p. 293. CCIS L2-245, Eocene 18, 02

UALVP39002 Liu *et al.*, 2016, fig. Fig. 3, p. 293. CCIS L2-245, Eocene 18, 02

UALVP39003 Liu *et al.*, 2016, fig. Fig. 3, p. 293. CCIS L2-245, Eocene 18, 02

UALVP39004 Liu *et al.*, 2016, fig. Fig. 3, p. 293. CCIS L2-245, Eocene 18, 02

UALVP39005 Liu *et al.*, 2016, fig. Fig. 3, p. 293. CCIS L2-245, Eocene 18, 02

UALVP39006 Liu *et al.*, 2016, fig. Fig. 3, p. 293. CCIS L2-245, Eocene 18, 02

UALVP39007 Liu *et al.*, 2016, fig. Fig. 3, p. 293. CCIS L2-245, Eocene 18, 02

UALVP39008 Liu *et al.*, 2016, fig. Fig. 3, p. 293. CCIS L2-245, Eocene 18, 03

UALVP39009 Liu *et al.*, 2016, fig. Fig. 3, p. 293. CCIS L2-245, Eocene 18, 02

UALVP39010 Liu *et al.*, 2016, fig. Fig. 3, p. 293. CCIS L2-245, Eocene 18, 03

UALVP39011 Liu *et al.*, 2016, fig. Fig. 3, p. 293. CCIS L2-245, Eocene 18, 03

UALVP39012 Liu *et al.*, 2016, fig. Fig. 3, p. 293. CCIS L2-245, Eocene 18, 03

UALVP39013 Liu *et al.*, 2016, fig. Fig. 3, p. 293. CCIS L2-245, Eocene 18, 03

UALVP39014 Liu *et al.*, 2016, fig. Fig. 3, p. 293. CCIS L2-245, Eocene 18, 03

UALVP39015 Liu *et al.*, 2016, fig. Fig. 3, p. 293. CCIS L2-245, Eocene 18, 03

UALVP39016 Liu *et al.*, 2016, fig. Fig. 3, p. 293. CCIS L2-245, Eocene 18, 03

UALVP39017 Liu *et al.*, 2016, fig. Fig. 3, p. 293. CCIS L2-245, Eocene 18, 03

UALVP39018 head missing. Liu *et al.*, 2016, fig. Fig. 3, p. 293. CCIS L2-245, Eocene 18, 03

UALVP39019 part & counterpart. Liu *et al.*, 2016, fig. Fig. 3, p. 293. CCIS L2-245, Eocene 18, 03

UALVP39020 on slab with UALVP39021. Liu *et al.*, 2016, fig. Fig. 3, p. 293. CCIS L2-245, Eocene 18, 03

UALVP39021 on slab with UALVP39020. Liu *et al.*, 2016, fig. Fig. 3, p. 293. CCIS L2-245, Eocene 18, 03

UALVP39022 Liu *et al.*, 2016, fig. Fig. 3, p. 293. CCIS L2-245, Eocene 18, 03

UALVP39023 Liu *et al.*, 2016, fig. Fig. 3, p. 293. CCIS L2-245, Eocene 18, 03

UALVP39024 part & counterpart; tail missing. Liu *et al.*, 2016, fig. Fig. 3, p. 293. CCIS L2-245, Eocene 18, 03

UALVP39025 part & counterpart. Liu *et al.*, 2016, fig. Fig. 3, p. 293

UALVP39026 Liu *et al.*, 2016, fig. Fig. 3, p. 293. CCIS L2-245, Eocene 18, 04

UALVP39027 Liu *et al.*, 2016, fig. Fig. 3, p. 293. CCIS L2-245, Eocene 18, 04

UALVP39028 Liu *et al.*, 2016, fig. Fig. 3, p. 293. CCIS L2-245, Eocene 18, 04

UALVP39029 Liu *et al.*, 2016, fig. Fig. 3, p. 293. CCIS L2-245, Eocene 18, 04

UALVP39030 Liu *et al.*, 2016, fig. Fig. 3, p. 293. CCIS L2-245, Eocene 18, 04

UALVP39031 Liu *et al.*, 2016, fig. Fig. 3, p. 293. CCIS L2-245, Eocene 18, 04

UALVP39032 on the same slab, UALVP39033, UALVP39034. Liu *et al.*, 2016, fig. Fig. 3, p. 293. CCIS L2-245, Eocene 18, 04

UALVP39033 on the same slab, UALVP39032, UALVP39034. Liu *et al.*, 2016, fig. Fig. 3, p. 293. CCIS L2-245, Eocene 18, 04

UALVP39034 on the same slab, UALVP39032, UALVP39033. Liu *et al.*, 2016, fig. Fig. 3, p. 293. CCIS L2-245, Eocene 18, 04

UALVP39035 Liu *et al.*, 2016, fig. Fig. 3, p. 293. CCIS L2-245, Eocene 18, 04

UALVP39036 part & counterpart. Liu *et al.*, 2016, fig. Fig. 3, p. 293. CCIS L2-245, Eocene 18, 08

UALVP39037 part & counterpart. Liu *et al.* 2016, fig. Fig. 3, p. 293. CCIS L2-245, Eocene 18, 08

UALVP39038 part & counterpart on slab with UALVP39039. Liu *et al.*, 2016, fig. Fig. 3, p. 293. CCIS L2-245, Eocene 18, 02

UALVP39039 part & counterpart on slab with UALVP39038. Liu *et al.*, 2016, fig. Fig. 3, p. 293. CCIS L2-245, Eocene 18, 02

UALVP52373 two pieces, a, b. Liu *et al.*, 2016, fig. Fig. 3, p. 293. CCIS L2-245, Eocene 21, 03

Osteichthyes Huxley, 1880
  Esociformes Bleeker, 1859
    Esocidae Cuvier, 1817
UALVP15002 HOLOTYPE *Esox tiemani* Wilson, 1980, p. 309
Collector: Tieman, Bert, 1978

Canada: Alberta, Smoky Tower, Smoky Tower #1, 54.4237°N. Lat., 118.295°W. Long. Tertiary-Paleocene, Paskapoo. Almost complete, part & counterpart, partially digested fish in gut, full body cast and head cast. Wilson, 1980, Descr. p. 309. & fig. Fig. 2, a–b, p. 310. Wilson, 1984, Palaeont., Vol. 27(3):597–608, fig. Fig. 1, p. 599, Fig. 2, a, b, p. 601, Fig. 5, p. 603, Fig. 6, p. 604, Wilson and Williams, 2010, Origin & Phylo. Interrel. of Teleosts, Descr. pp. 379–409, & fig. Fig. 1, A, p. 380. CCIS L2-245, Paleocene 06, 07

Wilson, Mark V. H. 1980. Oldest known *Esox* (Pisces: Esocidae), part of a new Paleocene teleost fauna from western Canada. Canadian Journal of Earth Sciences, Vol. 17(3):307–312

PARATYPES:
UALVP15005 skull. Wilson, 1980, Descr. p. 309, & fig. Fig. 2, c, p. 310. Wilson, 1984, Palaeont., Vol. 27(3):597–608. CCIS L2-245, Paleocene 06, 07

UALVP15006 near complete. Wilson, 1980, Descr. p. 309. Wilson, 1984, Palaeont., Vol. 27(3):597–608, fig. Fig. 3, p. 602. CCIS L2-245, Paleocene 06, 07
UALVP15070 lacking head & tail. Wilson, 1980, Descr. p. 309. Wilson, 1984, Palaeont., Vol. 27(3):597–608, fig. Fig. 4 p. 602. CCIS L2-245, Paleocene 06, 07
UALVP15071 partial. Wilson, 1980, Descr. p. 309. Wilson, 1984, Palaeont. Vol. 27(3):597–608, fig. Fig. p. 602. CCIS L2-245, Paleocene 06, 07
UALVP15072 partial. Wilson, 1980, Descr. p. 309. Wilson, 1984, Palaeont. Vol. 27(3):597–608, fig. Fig. 4, p. 60. CCIS L2-245, Paleocene 06, 07

Osteichthyes Huxley, 1880
  Esociformes Bleeker, 1859
    Umbridae Bleeker, 1859
UALVP50866 Cast of HOLOTYPE Cast of Holotype UMMP V57007, University of Michigan Museum of Paleontology. *Novumbra oregonensis* **Cavender, 1969, p. 5**
Collector: Edward Frazer, July 1964
The United States: Oregon: Wheeler County: nine miles northwest of Mitchell, NE1/4, Sec. 10, T11S, R20E, Knox Ranch. Tertiary-Oligocene-middle (Orellan), John Day Formation Complete specimen except for caudal fin, part & counterpart. Cavender, 1969, Descr. pp, 2, 5, 7, 8, 9, 11, & fig. Fig. 1, p. 6., Plate 1, p. 31

    Cavender, Ted 1969 An Oligocene mudminnow (Family Umbridae) from Oregon with remarks on relationships within the Esocoidei. Occasional Papers of the Museum of Zoology University of Michigan Number 660:1–33
<u>PARATYPE:</u>
UALVP13404 casts of Paratype UMMP V57008 Complete specimen, Cavender, 1969, Descr. p. 11, & fig. Fig. 3, B, p. 10 (V57008); and Paratype UMMP V57004 Disarticulated skull. Cavender, 1969, Descr. p. 2, & fig. Fig. 2, B, p. 8 (V57004)

UALVP50867 cast of Paratype UMMP V57008 nearly complete specimen part of axial skeleton, part & counterpart. Cavender, 1969, Descr. pp, 2, 7, 8, 11, & fig. Fig. 3, B, p. 10, Fig. 4, p. 12.

Osteichthyes Huxley, 1880
  Osmeriformes
    Osmeridae Regan, 1913
UALVP31703 HOLOTYPE *Speirsaenigma lindoei* Wilson and Williams, 1991
Collector: Lindoe, L. Allan, 1989
Canada: Alberta, Joffre, Joffre Bridge SW Fish Layer. Tertiary-Paleocene, Paskapoo Formation. complete, part & counterpart, 60 mm., 29.06, 5.06, 230! Wilson and Williams, 1991, Descr. p. 436, & fig. Fig. 2, p. 436, caudal skeleton Fig. 9, p. 441; Wilson and Williams, 2010, Origin & Phylo. Interrel. of Teleosts 379–409. fig. Fig. 2, C, p. 384. CCIS L2-245, Paleocene 08, 10
  Wilson, Mark V. H., and Williams, Robert R. G. 1991. New Paleocene Genus and Species of Smelt (Teleostei:Osmeridae) from freshwater deposits of the Paskapoo Formation, Alberta, Canada, and comments on Osmerid Phylogeny, Journal of Vertebrate Paleontology, Vol. 11(4):434–451
PARATYPES:
UALVP21529 skull & partial body, part & counterpart. Wilson and Williams, 1991, Descr. p. 436, p. 442
UALVP21532 skull, part & counterpart. Wilson and Williams, 1991, Descr. p. 436
UALVP21533 most of the specimen. Wilson and Williams, 1991, Descr. p. 436, mesopterygoid p. 442, ceratobranchial, supracleithrum p. 443
UALVP21534 skull & partial, part & counterpart. Wilson and Williams, 1991, Descr. p. 436, & fig. Fig. 3, p. 437, proethmoids, p. 439, parasphenoid, p. 440, prootic, supraoccipital p. 441, anguloarticular p. 442, ceratohyal, supracleithrum p. 443, pectoral radials, pelvic fin and girdle p. 444. Biological Sciences Building Z425, Unit 13, Drawer 1

UALVP21535 partial, part & counterpart. Wilson and Williams, 1991, Descr. p. 436. anguloarticular, retroarticular p. 442

UALVP21536 partial. Wilson and Williams, 1991, Descr. p. 436, male with enlarged anal fin with thickened rays and pterygiophores p. 444

UALVP21537 partial. Wilson and Williams, 1991, Descr. p. 436, presumed male, p 444

UALVP21540 partial, part & counterpart. Wilson and Williams, 1991, Descr. p. 436, & fig. caudal skeleton Fig. 8, p. 440; Fig. 9, p. 441. Male p. 444

UALVP21541 almost complete, part & counterpart, Tables 1, 2, p. 431. Wilson and Williams, 1991, fig. Fig. 4, p. 438, & Descr. proethmoids, lateral ethmoid (or, orbitosphenoid) p. 439, parietal, parasphenoid p. 440, sphenotic, pterotic, prootic, supraoccipital p. 441, ceratohyal, interoperculum, 5th ceratobranchial, supracleithrum p. 443, pectoral radials, pelvic fin & girdle p. 444 paratype of *Speirsaenigma lindoei*. Murray and Wilson, 1999, published as *Lateopisculus turrifumosus* p. 400. CCIS L2-245, Paleocene 08, 10

UALVP21545 skull & partial body, part & counterpart. Wilson and Williams, 1991, Descr. p. 436, & fig. skull showing otoliths Fig. 5, p. 439, ectopterygoid Fig. 7 A, mesopterygoid Fig. 7 B, p. 440, prootic, p. 441, premaxilla p. 441, mesopterygoid p. 442, ceratohyal p. 443

UALVP23479 complete. 85-1-G. Wilson and Williams, 1991, Descr. p. 436, presumed male p. 444. CCIS L2-245, Paleocene 08, 10

UALVP23480 disarticulated part & counterpart. 85-1-T1. Wilson and Williams, 1991, Descr. p. 436

UALVP23481 partial. 85-1-P2. Wilson and Williams, 1991, Descr. p. 436

UALVP23482 partial. 85-1-G. Wilson and Williams, 1991, Descr. p. 436

UALVP23483 partially disarticulated part & counterpart. Wilson and Williams, 1991, Descr. p. 436, supraorbital p. 439

UALVP23484 three specimens, partial & complete part & counterpart. 85-1-ZX. Wilson and Williams, 1991, Descr. p. 436, parasphenoid p. 440

UALVP23485 minus head part & counterpart. 85-1-ZZ. Wilson and Williams, 1991, Descr. p. 436, female, p. 444

UALVP23486 partial. 85-1-G. Wilson and Williams, 1991, Descr. p. 436

UALVP23487 head part & counterpart. 85-1-ZY. Wilson and Williams, 1991, Descr. p. 436, parietal, parasphenoid p. 440, premaxilla p. 441, anguloarticular p. 442

UALVP23488 one complete & one tail. Roughly in FF0 or GG0. Wilson and Williams, 1991, Descr. p. 436, & fig. urohyal Fig. 7, C, p. 440, caudal skeleton Fig. 9, p. 441, urohyal p. 443, pelvic fin & girdle, anal fin p. 444, 10 to 12 predorsals, female, caudal skeleton & fin p. 444. CCIS L2-245, Paleocene 08, 10

UALVP26051 part & counterpart; P.L.: 24.09; P.I. 4.3; 156!; field no. X-4-1. Wilson and Williams, 1991, Descr. p. 436

UALVP30830 imperfect complete, part & counterpart, 32.23, 1.04, 276D. Wilson and Williams, 1991, Descr. p. 436

UALVP30835 disarticulated bones, part & counterpart, 32.52, 1.28. Wilson and Williams, 1991, Descr. p. 436

UALVP30853 part of head missing, part & counterpart, female, 47 mm, 32.00, 2.72, 298D. Wilson and Williams, 1991, Descr. p. 436

UALVP30867 complete, tail damaged, part & counterpart, 52 mm, 31.18, 2.81, 291D. Wilson and Williams, 1991, Descr. p. 436

UALVP30872 complete, part & counterpart, 53 mm, 31.95, 2.80, 259D. Wilson and Williams, 1991, Descr. p. 436

UALVP30873 complete, part & counterpart, 58 mm, 31.30, 2.90, 68D. Wilson and Williams, 1991, Descr. p. 436

UALVP30874 skull & disarticulated body, part & counterpart, 31.39, 2.85, 6D. Wilson and Williams, 1991, Descr. p. 436

UALVP30882 complete, part & counterpart, 46.8 mm, 30.15, 2.65, 164D. Wilson and Williams, 1991, Descr. p. 436

UALVP30887 partial, part & counterpart, 30.64, 2.73, 289D. Wilson and Williams, 1991, Descr. p. 436

UALVP30892 partial, tail missing, two pieces, 30.73, 3.50, 104D. Wilson and Williams, 1991, Descr. p. 436

UALVP30897 partial, two pieces, 30.43, 3.65. Wilson and Williams, 1991, Descr. p. 436

UALVP30898 partial, two pieces, 30.43, 3.69. Wilson and Williams, 1991, Descr. p. 436

UALVP30900 posterior half, 31.82, 3.22, 31D. Wilson and Williams, 1991, Descr. p. 436

UALVP30924 imperfect, complete, part & counterpart, 21.51, 3.60. Wilson and Williams, 1991, Descr. p. 436

UALVP30925 complete, part & counterpart, male, 42.9 mm., 21.64, 3.64, 104D. Wilson and Williams, 1991, Descr. p. 436, & fig. caudal skeleton Fig. 9, p. 441, caudal skeleton p. 444

UALVP30928 partial, three pieces, 21.69, 3.55, 180D. Wilson & Williams, 1991, Descr. p. 436, presumed male, supraneurals p. 444

UALVP31642 head missing, part & counterpart, 25.06, 4.50, 21! Wilson & Williams, 1991, Descr. p. 436, presumed male p. 444. CCIS L2-245, Paleocene 08, 10

UALVP31679 complete, male, 63 mm, 29.33, 3.91, 239! Wilson & Williams, 1991, Descr. p. 436, dentary p. 442, presumed male, anal fin with 20 pterygiophores p. 444. CCIS L2-245, Paleocene 08, 10

UALVP31682 tail missing, 29.41, 3.95, 292! Wilson and Williams, 1991, Descr. p. 439, vomer, supraoccipital p. 441. CCIS L2-245, Paleocene 08, 10

UALVP31688 complete, 29.79, 4.19, 0! Wilson and Williams, 1991, Descr. p. 439. CCIS L2-245, Paleocene 08, 10

UALVP31692 imperfect, complete, 29.87, 4.30, 13! Wilson and Williams, 1991, Descr. p. 439. CCIS L2-245, Paleocene 08, 10

UALVP31723 (on the same rock as UALVP30853). Wilson and Williams, 1991, Descr. p. 439

UALVP31737 skull only, 23.31, 4.16, 264D, W-4–8. Wilson and Williams, 1991, Descr. p. 439. CCIS L2-245, Paleocene 08, 10

Literature Cited

Berry, W. B. N. 1985. Chapter 3. The significance of type specimens and old collections to research in the biological sciences. pp. 23–37. *IN:* Miller, E. H. (editor). Museum Collections: Their roles and future in Biological Research. Occasional Papers of the British Columbia Provincial Museum No. 25:1–221.

Bruner, John Clay. 1992. A catalogue of type specimens of fossil fishes in the Field Museum of Natural History Fieldiana: Geology New Series. No. 23(Publication 1431):1–54

Cappetta, Henri. 1987. Chondrichthyes II. Mesozoic and Cenozoic Elasmobranchii Volume 3B pp. 1–193. *IN:* Schultze, Hans-Peter (editor) Handbook of Paleoichthyology. Gustav Fischer Verlag, Stuttgart, Germany.

Cavender, T. M. 1966. Systematic position of the North American Eocene fish, *"Leuciscus" rosei* Husakof. Copeia 1966(2):311–320.

Denison, Robert. 1978. Placodermi Volume 2. pp. 1–128. *IN:* Schultze, Hans-Peter (editor) Handbook of Paleoichthyology. Gustav Fischer Verlag, Stuttgart, Germany.

Grande, Lance, and Bemis, William E. 1998. A comprehensive phylogenetic study of Amiid fishes (Amiidae) based on comparative skeletal anatomy. An empirical search for interconnected patterns of Natural History. Society of Vertebrate Paleontology Memoir 4:1–690.

Langston, Wann, Jr. (Chairman), Black, Craig C., Lillegraven, Jason A., Patton, Thomas C., Wilson, Robert W., and Schaeffer, Bobb. 1977. Fossil Vertebrates in the United States. The Next Ten Years. Report of the Society of Vertebrate Paleontology Advisory Committee for Systematics Resources in Vertebrate Paleontology. Society of Vertebrate Paleontology. Austin, Texas. 40 pp.

Märss, Tiuu, Turner, Susan, and Karatajute-Talimaa, Valentina. 2007. "Agnatha" II Thelodonti. Volume 1B. pp. 1–143. *IN:* Schultze, Hans-Peter (editor) Handbook of Paleoichthyology. Verlag Dr. Friedrich Pfeil, München, Germany.

Nelson, Joseph S., Grande, Terry C., and Wilson, Mark V. H. 2016. Fishes of the World. 5th Edition. John Wiley & Sons, Inc., Hoboken, New Jersey, U.S.A., 752 pp.

Zangerl, Rainer. 1981. Chondrichthyes I. Paleozoic Elasmobranchii. Volume 3A. pp. 1–115. *IN:* Schultze, Hans-Peter (editor) Handbook of Paleoichthyology. Gustav Fischer Verlag, Stuttgart, Germany.

Systematic Index

Printed and bound by CPI Group (UK) Ltd, Croydon, CR0 4YY

23/10/2024

01778243-0005